GW01221091

MACK TRUCKS

HENRY RASMUSSEN

MBI Publishing Company

First published in 2001 by MBI Publishing Company, Galtier Plaza, Suite 200, 380 Jackson Street, St. Paul, MN 55101-3885 USA

© Rasmussen, Henry, 1987, 2001

All rights reserved. With the exception of quoting brief passages for the purposes of review, no part of this publication may be reproduced without prior written permission from the Publisher.

The information in this book is true and complete to the best of our knowledge. All recommendations are made without any guarantee on the part of the author or Publisher, who also disclaim any liability incurred in connection with the use of this data or specific details.

We recognize that some words, model names and designations, for example, mentioned herein are the property of the trademark holder. We use them for identification purposes only. This is not an official publication.

MBI Publishing Company books are also available at discounts in bulk quantity for industrial or sales-promotional use. For details write to Special Sales Manager at Motorbooks International Wholesalers & Distributors, Galtier Plaza, Suite 200, 380 Jackson Street, St. Paul, MN 55101-3885 USA.

Library of Congress Cataloging-in-Publication Data Available

ISBN 0-7603-1218-4

Printed in Hong Kong

On the front cover: Mack's classic mascot, in this case on the nose of a 1940 model EH.

On the frontispiece: Fifties-era radiator emblem. Pitted chrome, chipped paint, and a graveyard of dead insects indicate a rig still hard at work.

On the title page: This 1980s Super-Liner dump truck at speed is typical of hard-working Macks.

On the table of contents page: Who better to stand guard over a classic Mack? John Hulse of Seattle, Washington, nevers worries when leaving his 1953 LTL unattended.

On the back cover: At left is a 1980s era Mid-Liner, in this case a CS model. Power comes from a Renault-built, 335-cubic-inch, diesel six. To the right is an Ultra-Liner, introduced in 1983.

Contents

Introduction
"Keep'em comin', Mack!". 6
Mack Classic
The bulldog conquers the world. 8
Mack Man
Amassing memorable machinery. 16
Big Mack
The symbols of power and prestige. 22
Family Mack
When Mack runs in the blood. 38
Fifties' Mack
Bold faces of a boom era. 48
Mack Woman
She is the "Lumber Gal". 58
Working Mack
Hauling the loads of the land. 68
Modern Macks
High-tech raiders of the road. 102
Mack Mania
Truck buffs show and tell. 112

Introduction

"Keep'em comin', Mack!"

Few products in the technological boom of our century inspires as much nostalgic patriotism as does Mack. There are others to be sure. Names like Jeep, Harley-Davidson, Caterpillar, come to mind. They stand on their own, symbolizing their own accomplishments, such as off-road mobility, two-wheeled excitement, brute power. But Mack combines some of them all, the mobility, the excitement, the power—and adds that awesome, long-haul durability.

And then there is the name itself. Tasting so good on the tongue. Right smack in the middle. Sounding so strong to the ear. Like the crack of a shotgun. And there are the phrases. In old books. In old movies. *Hey, Mack! Have a match? You alright, Mack? Keep'em comin', Mack!*

And there are the slogans. *Performance Counts. If You've Got a Mack, You're Lucky. If You Plan To Get One, You're Wise. Modernize with Mack.* And then there is the other phrase. *Built like a Mack.* The phrase that never had to be advertised. Because it was coined by the people.

John Mack, it was he who started it all. Born in 1864, near Mount Cobb, Pennsylvania, he was the middle son of five boys born to a German emigree. John ran away from home at the age of fourteen. Still on the run, he found employment first as the driver of a mule cart, then as a fireman, then as the operator of a hoisting engine. Later he became the engineer on a ship sailing around Central America. After spending some time in the Caribbean, he finally returned to the USA, arriving in New York in 1890.

John this time found work as the engineer of a stationary steam engine in the Brooklyn wagon factory where his brother Augustus was a clerk. Three years later, when one of the partners retired, the brothers took over the company, John being the driving force in the venture. The following year, they were joined by a third brother, William. For almost a decade the brothers kept producing wagons, but in 1903 they decided to go horseless, constructing a sightseeing bus. In 1904, a fourth brother, Joseph, joined the firm. The same year, the Mack Brothers began building their own engines, and as orders for the buses and their new product—the trucks—began streaming in, new facilities were needed, and a move to Allentown, Pennsylvania, was organized in 1905.

And Allentown is still the site of the Mack headquarters. But the Mack Brothers are out of the picture. Have been for a long time. Already in 1912, when additional capital was needed, the brothers had become involved with Wall Street investors, who soon bought them out. John tried his hand once more at building trucks. The result was the Maccar, which remained as an independent marque until the mid-thirties. But long before, John had left the company to operate his own truck agency in Allentown. In 1924, an automobile accident took his life—he did not live to see the ultimate success of the company he had founded.

Today, Mack builds trucks in Allentown and Macungie, Pennsylvania, in Hagerstown, Maryland, and in Winnsboro, South Carolina, where a new state-of-the-art facility has just been opened. Mack also operates factories in Canada and Australia. Total production in 1986 was approximately 20,000 units. When the overseas climate was more favorable than it is today—in the mid-seventies—the annual production reached the 30,000 level. During that period Mack employed approximately 14,000 workers. Today, that number is in the neighborhood of 10,000. But a leaner operation means a more efficient operation, and Mack is well set for an exciting future.

In November of 1986, Mack celebrated the completion of its one millionth vehicle. America's oldest and most prestigious truck manufacturer is indeed to be congratulated! *Keep'em comin', Mack!*

Mack Classic

The bulldog conquers the world

What Ford's Model T meant to the world of automobiles, Mack's model AC meant to the world of trucks. This milestone, with its characteristic hood design, was the brainchild of E.R. Hewitt. The work was begun while the gifted inventor was still the chief engineer of Mack, but reached its completion first under the reign of his successor, the equally gifted A.F. Masury. The first experimental units took to the road in 1915. Word of its rugged construction and awesome pulling power spread quickly across the nation. The ongoing war in Europe, and related orders from the military, meant that its fame soon reached the rest of the world. The AC was not marketed right away as the "Bulldog". It took the British—who immediately recognized the similarities to their fierce animal friend—to give it the famous nickname. Production began in 1916 and lasted until 1938, with more than 40,000 units manufactured—ample proof of the incredible impact of the Bulldog.

This 1916 survivor is the third oldest AC known to exist. The 3 1/2 ton workhorse was originally employed in the quarries of the Estoria Crushed Rock Company, Estoria, Oregon. Later it served as a water truck, also in Oregon. It was photographed—as were the examples pictured on the following six pages—in a collector's yard in Campo, a small town near California's border to Mexico. The early ACs had their hood openings covered by a mesh, as seen here, while subsequent units used louvers.

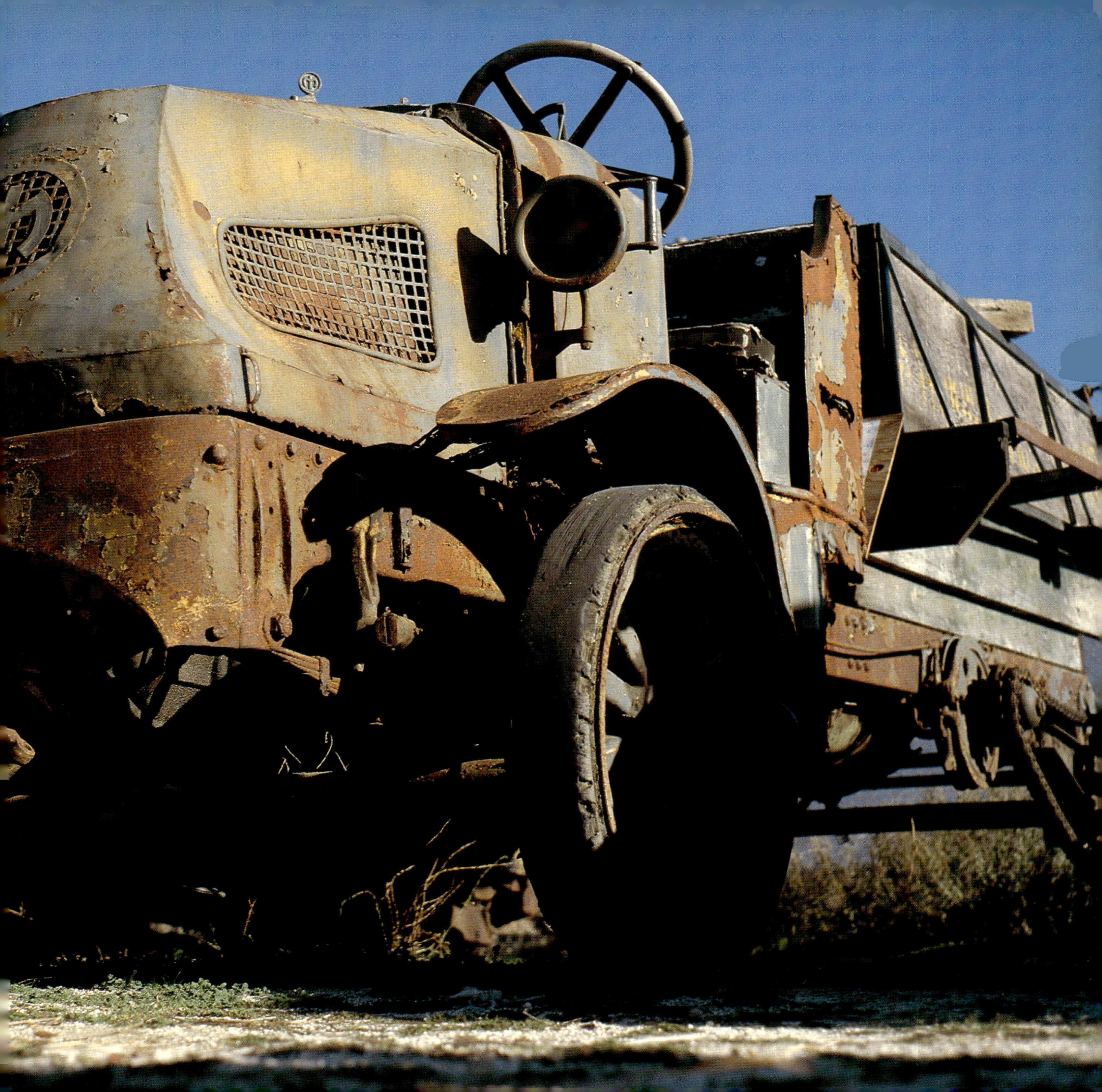

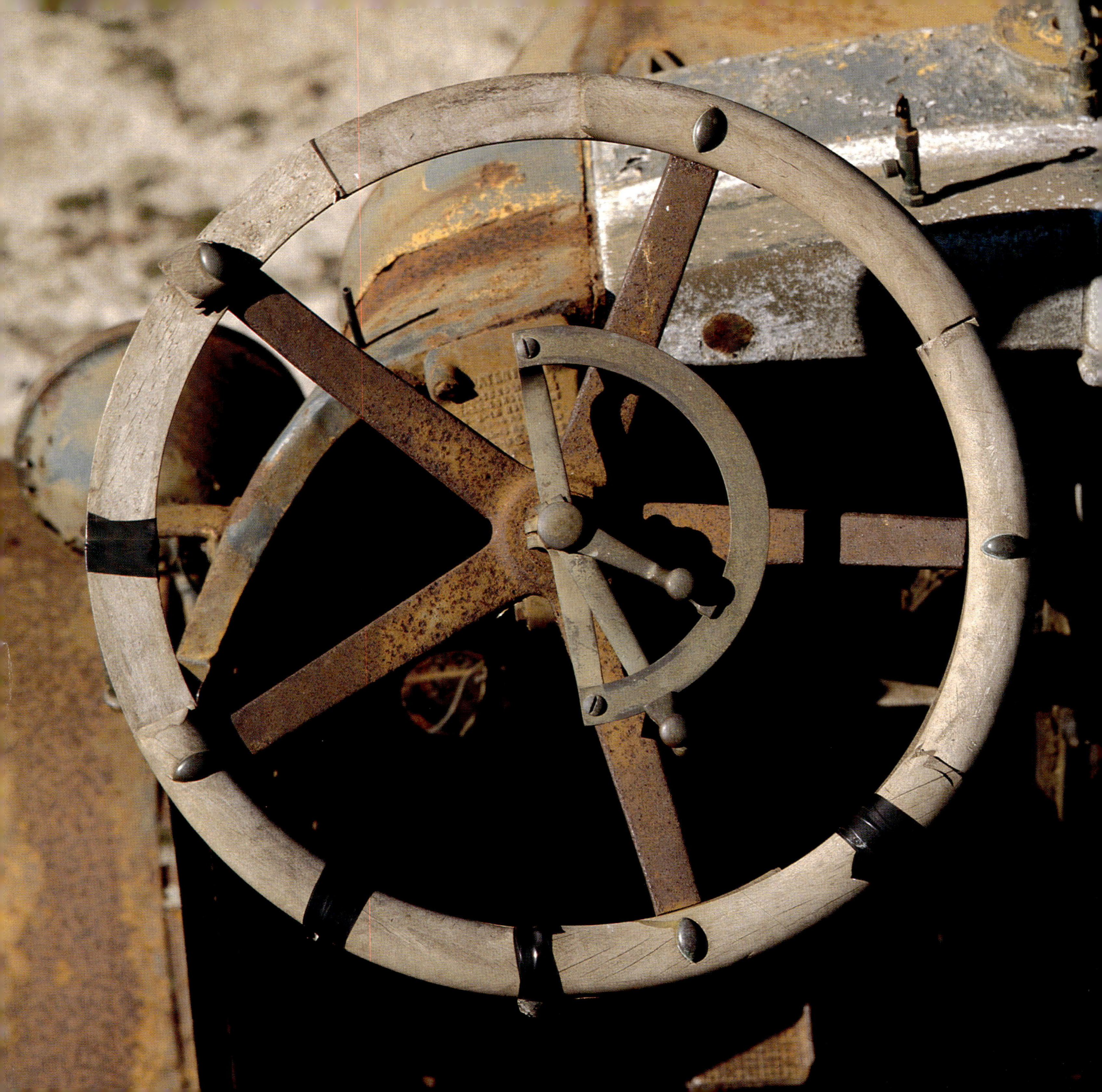

Pictured to the left, the classic five-spoke steering wheel of the 1916 AC, with its center-mounted ignition spark and throttle controls. The transmission was a three-speed affair, while the clutch was of a brake type, which meant that the driven member was prevented from spinning when the clutch was released for shifting. This feature proved a blessing, since the driver of the day was often quite careless in his attitude toward machinery.

Power was transmitted to the rear wheels by means of chain drive, as seen in the photograph above. Although quite noisy—an AC could always be distinguished by that certain whining sound from its chains—this feature was thought to have superior pulling characteristics. Another advantage was the higher rear axle clearance. The pros and cons of chain drive were hotly disputed, but its popularity is evidenced by the fact that Mack offered the feature as late as 1950. The early ACs rolled on wood-spoke wheels and solid rubber tires.

Next page
An impressive row of ACs, like the one pictured on the following spread, while commonly seen in the twenties and thirties, is an extremely rare sight today—perhaps found nowhere else in the world. Of course, in the old days, Mack trucks all came painted green. Curiously enough, when the door of one of the trucks in this line-up was raised— the doors slide up, rather than swing out—its steel surface was found to still carry the original lead-varnish Mack-green, still as shiny as new.

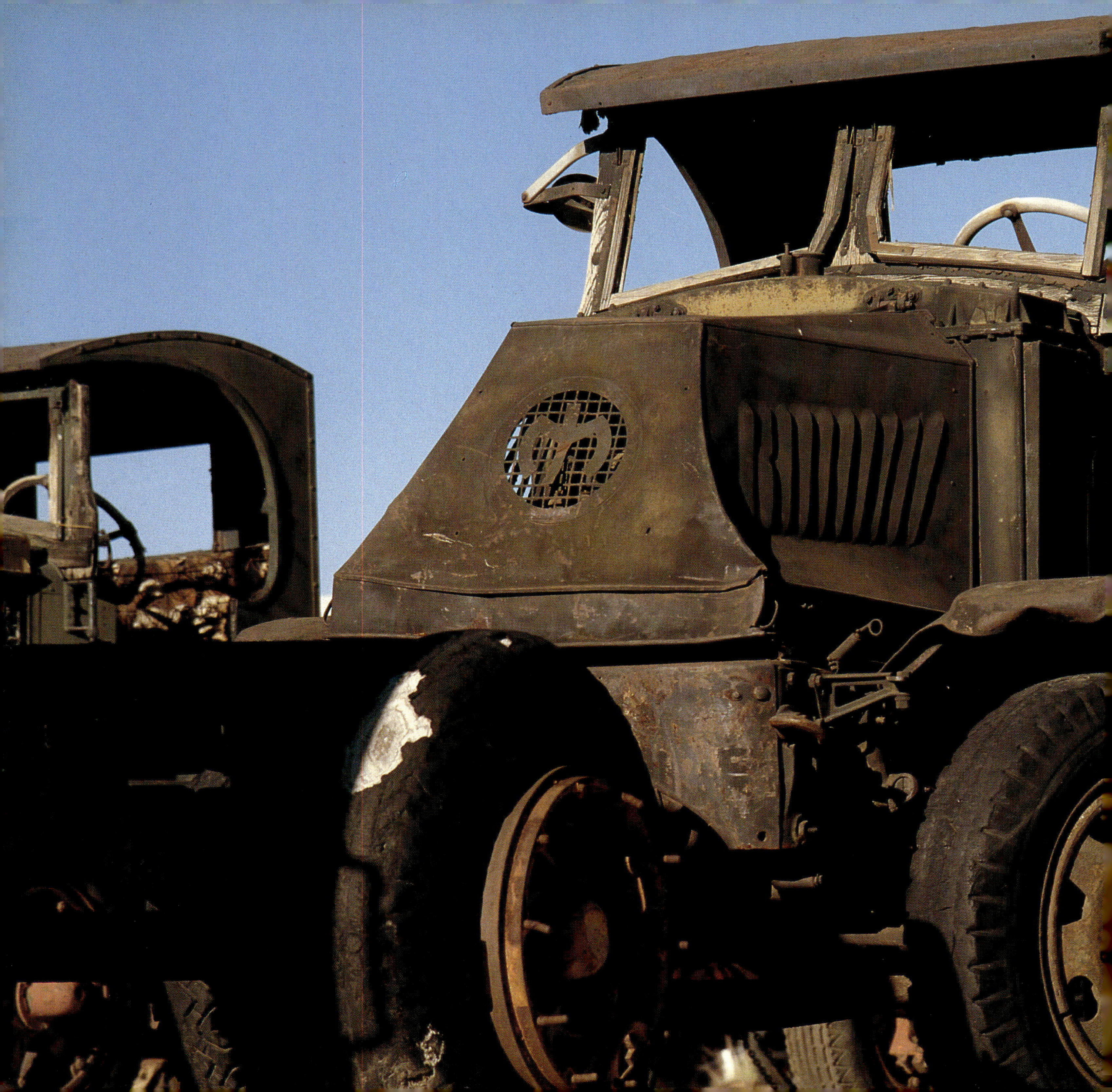

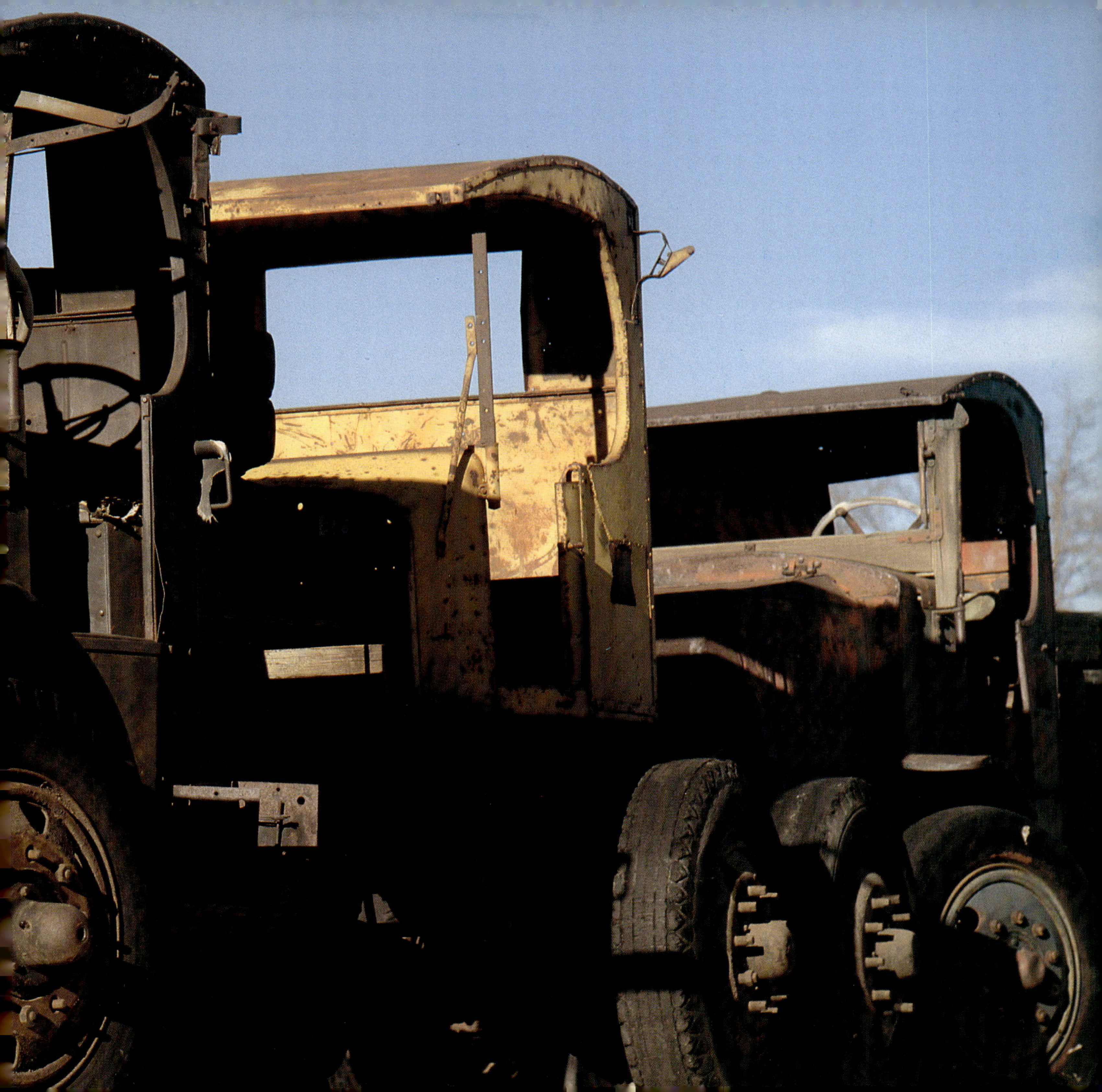

The AC was powered by a 4-cylinder in-line unit, constructed by Mack's own engineers and manufactured in-house. The bore was 5 inches and the stroke 6 inches. Displacement was 471 cubic inches. Output was between 40 and 50 bhp at 1,000 rpm. By the late thirties, power had risen to 74 bhp at 1,600 rpm. In the photograph above, with the right-hand side of the engine facing the camera, the exhaust manifold and its decorative cooling fins produce a most pleasing picture.

Pictured to the right, an AC of 1930 vintage. The radiator, as was the case during the entire production run, was located behind the engine, but it had already in 1922 become deeper and wider, extending several inches beyond the hood. The unique location of the radiator is said to have come about because drivers of horse-drawn wagons tended to back into this vulnerable component of early trucks—a way of showing their displeasure with the new order. Pneumatic tires did not become standard until 1934, but owners of out-of-date models often exchanged the old solid-rubbers as the new became available.

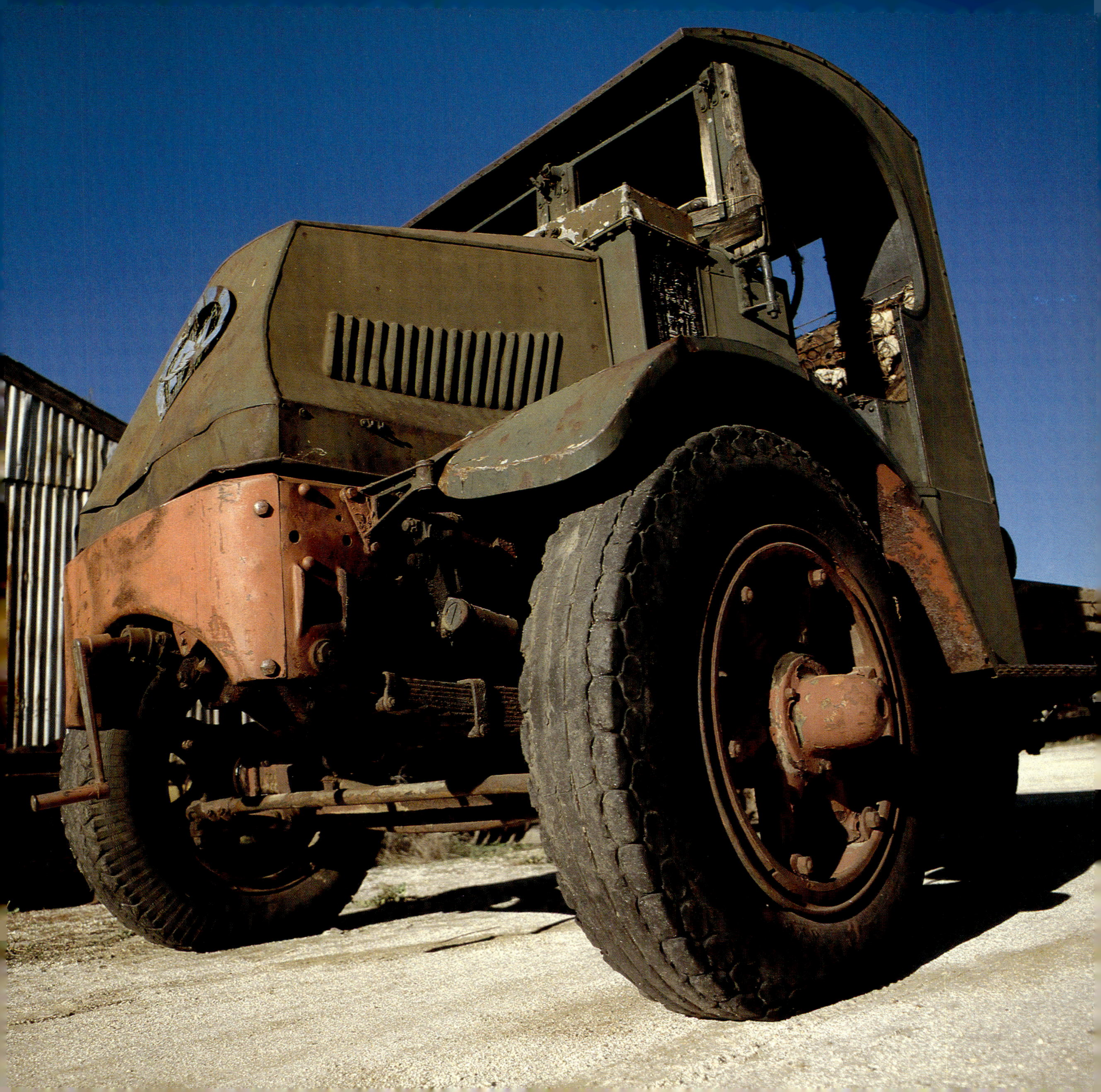

Mack Man

Amassing memorable machinery

Carl Calvert owns a real-estate business in Solana Beach, California. That is his profession. But his hobby is old Macks. Old Bulldog Macks, to be specific. He has more than a dozen of them now, collected from all over the nation—one of them was even brought back from as far away as Europe. This particular example is a 1918 AC, first shipped over to France, where it served its tour of duty with the US Army Corps of Engineers. After the hostilities had ended the AC was converted to civilian use and purchased by a Danish oil company. Decades later the faithful old Bulldog was brought to England by a Mack enthusiast. The restoration was performed by a subsequent owner. Calvert learned of the exquisite machine in 1985, corresponded with the owner and finally bought it sight unseen. It was driven onto a ship in Antwerp, Belgium, and arrived in Long Beach one month later. The rare old truck had spent the entire sea voyage on deck, a curious sight among its modern counterparts. No wonder, with its long and varied history, this particular Bulldog is Calvert's favorite.

This is how the Bulldog is cranked up. And Carl Calvert does it expertly by now. The photograph offers a good view of the left-hand side of the engine. Notice that the cylinders are cast in pairs of two. Notice also the location of the radiator, and the fact that it is covered by a mesh, while on later models—from 1922 on—it is covered with louvers.

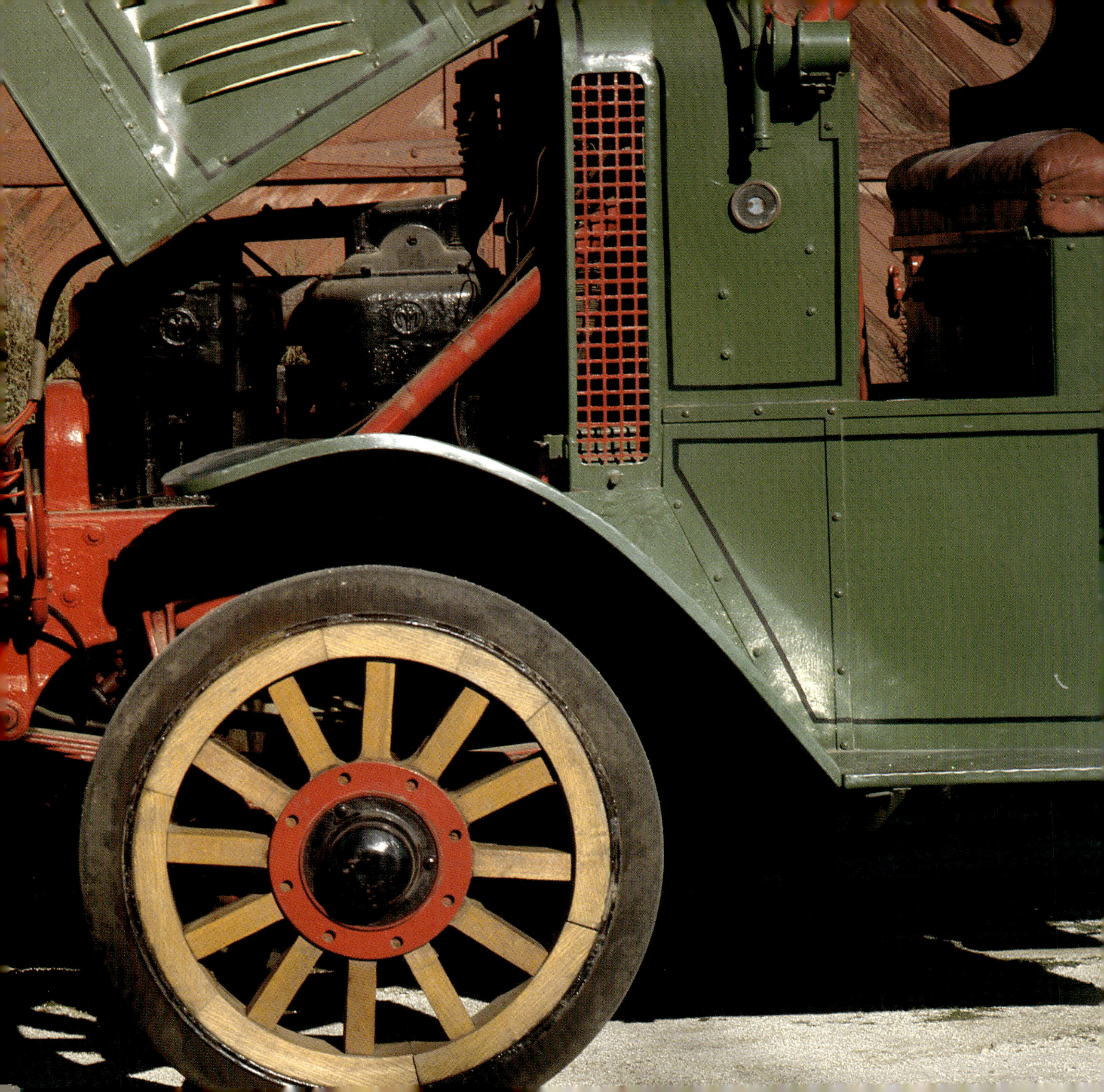

Above, a view of the Bulldog cockpit. The steering wheel is the original five-spoke unit, while the two white knobs—the spark and throttle controls—are subsequent modifications. Almost hidden in the deep shadow of the floor boards one can see the vague outline of two levers, one manipulating the gears, the other operating the box bed, which tips both to the rear and to the right. Not visible are the pedals—three of them—the clutch to the left, the brake to the right and the throttle in the middle.

In the photograph to the right, beams from the late sun illuminate the classic Mack logo, here—in all its gilded glory—somewhat prettier than it appeared in the old days. The picture also offers a view of the extremely rugged frame of the Bulldog, in particular its distinctively rounded front end, which also served as a heavy-duty bumper. Suspension was by leaf springs, their numbers depending on loading capacity.

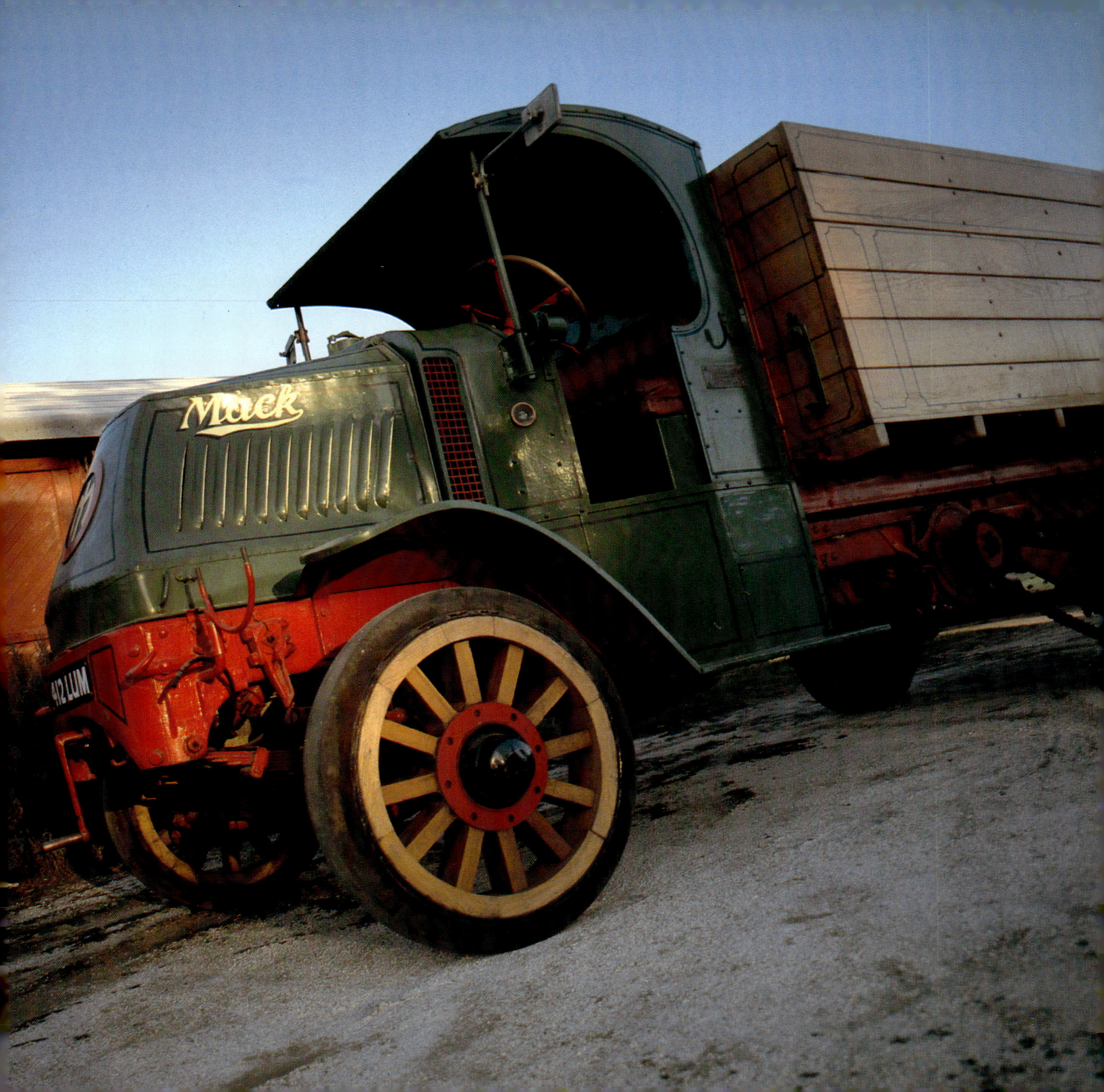
Mack
412 LUM

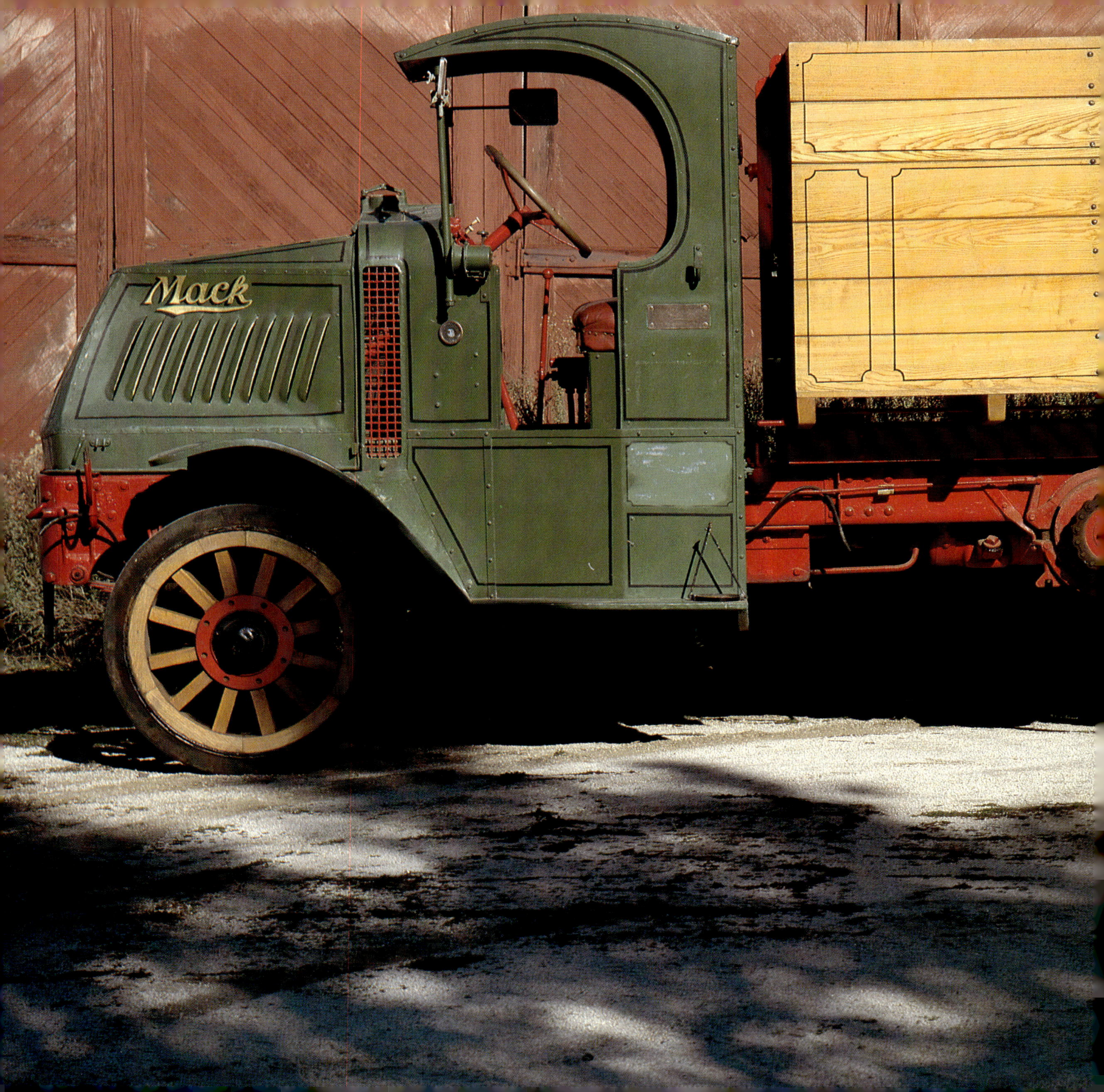
Mack

This side view shows Carl Calvert's famous Bulldog in all its compelling beauty. The box bed is most likely not of the original configuration; when used by the US Army the AC was normally fitted with the flat-bed type. The Bulldog came equipped with brakes only on the rear wheels. The emergency brakes are located at each end of the wheel axle, while the main brakes are incorporated into the jackshafts. The rod connecting to the pedal can be seen running forward along the length of the frame.

Big Mack

The symbol of power and prestige

Vehicles of particular prestige always seem to have special symbols connected with them, often carried proudly atop their radiators. Rolls-Royce has its winged lady; Mack has its bulldog. The pride of the Mack owner has been assigned a variety of characteristics and poses; it can look as if ready to attack, awesome with its stern gaze and fiercely forward-jutting jaw; it can look as if pulling an enormous load, aggressively braced, with muscles flexed. The bulldog in the form of a statue never adorned Mack's famous AC truck. But a bulldog pictured tearing to shreds a book bearing the title Hauling Costs was incorporated as a part of a name plate affixed to later AC models. The actual sculpture of the bulldog was created in 1932. The artist was Mack's versatile chief engineer, A.F. Masury. Several attempts were made to "improve" the original design, first in the form of a more rounded unit, in 1941, then as a comically stylized figure, in 1945, but nothing could replace the original, with its strong, angular lines—a classic, right from the beginning. For Mack it isn't the bulldog, it's just bulldog.

Pictured here is the legendary bulldog mascot as found on a 1940 model EH. In the initial version, as cast in 1932, the bulldog stood with its hind legs directly on the radiator cap (the artist's initials could be found on the left rear foot). Later, as the cap was moved under the hood, the bulldog was placed on a pedestal or a stand by itself.

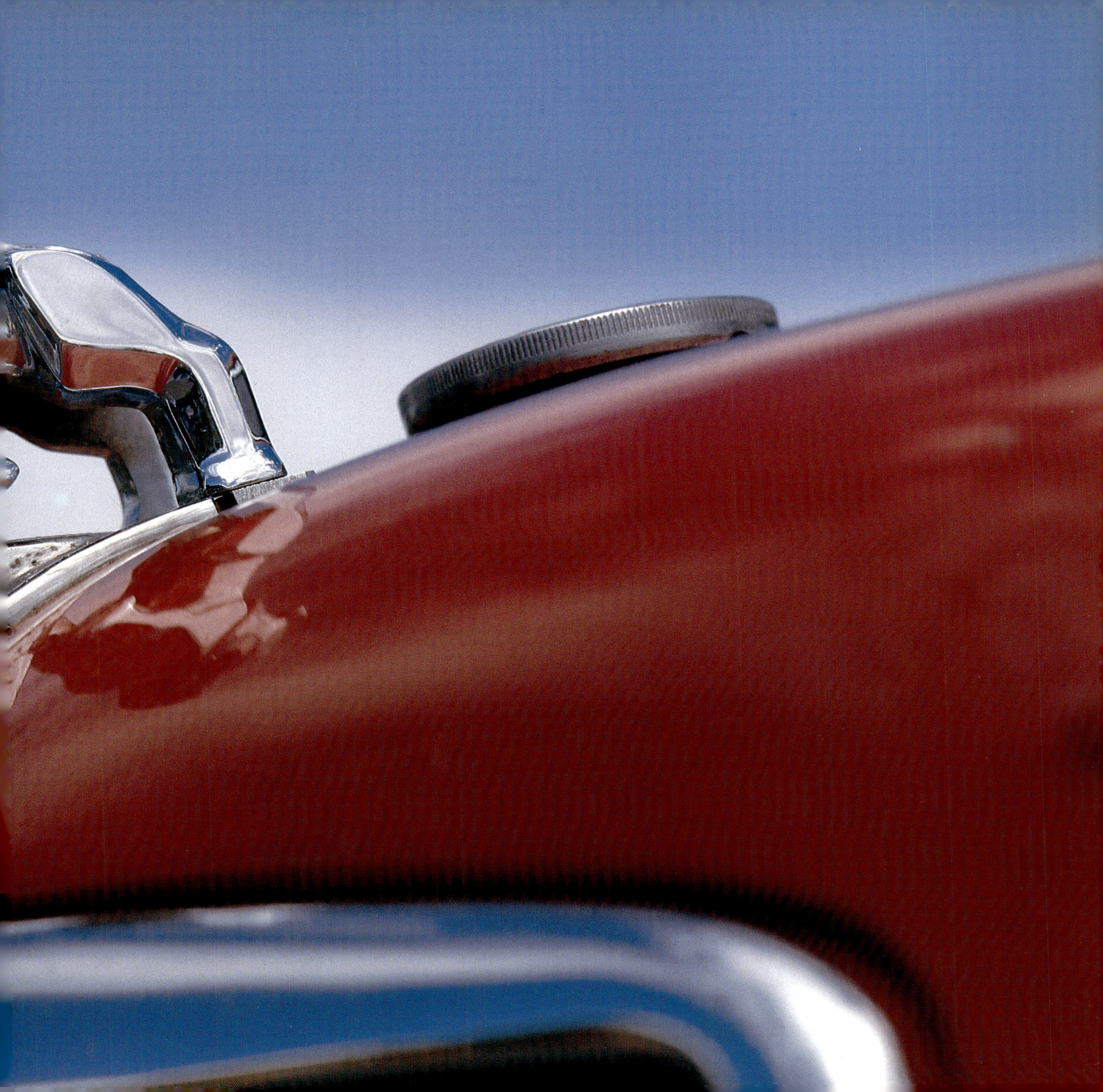

Seen in the picture above, the symbol that adorned the nose of every AC truck built, from the beginning in 1916 right up to the end in 1938. The original AC prototype did not carry the design, but photographs of a subsequent prototype shows it to have been applied very early on. The logo also appears on a variety of castings such as manifolds and transmission housings. Here the symbol has been placed on a non-original mesh. Also, somewhere along the line a portion of the ascendant has fallen off.

Pictured to the right, the bulldog mascot as applied to the side of the hood of a fifties' Mack. The features of the beast have become decidedly rounder in the transfer of the design, and the collar has received vicious-looking, sharply pointed spikes. Through the years, although pitted and ungroomed, the bulldog never seems to lose its poise—it comes across as fierce as always.

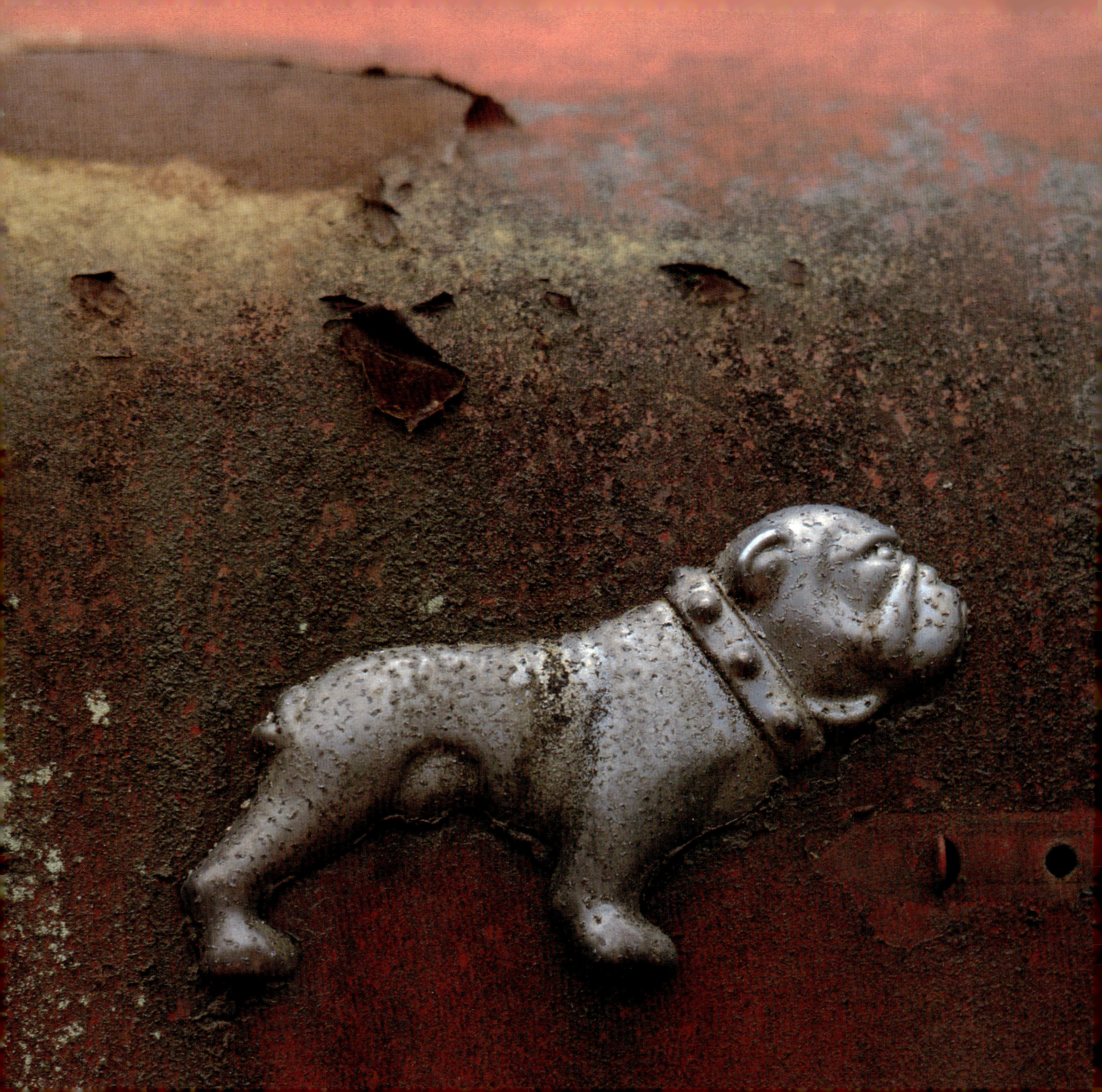

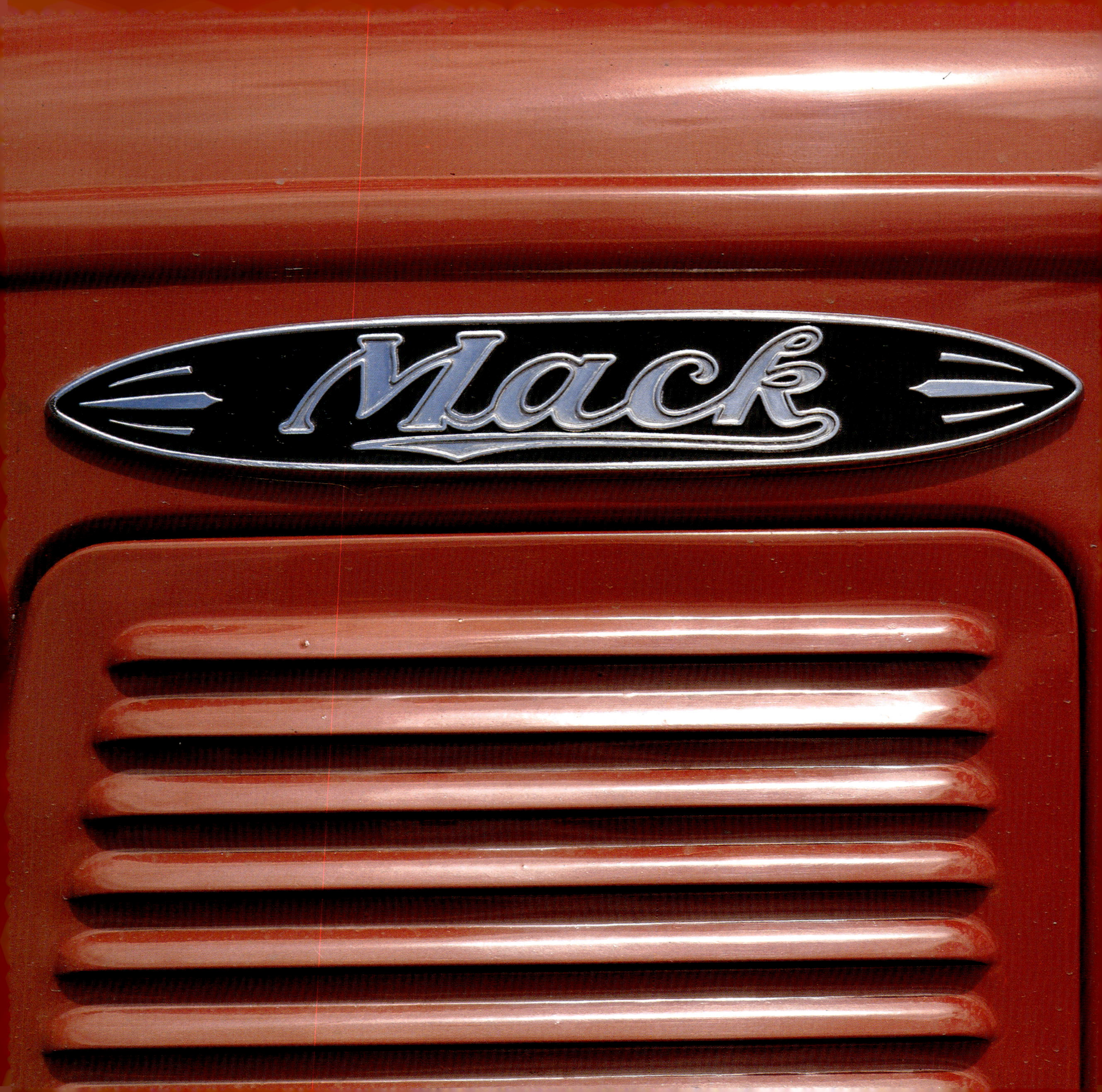
Mack

Featured on this spread are the emblems and markings of two beautifully restored Macks. To the left, the Mack insignia as found on a model EH, introduced in 1936. The design of the Mack script, with its sweeping tail—as was the case with many trademarks of the day—can be said to have its roots in the personal signature of the originator, that of John Mack. The design first appeared in 1910, at which time the trade name "Manhattan", which had been used by Mack up until that point, was dropped.

Shown above is a close-up side-view of a new line of Mack trucks, introduced in 1950, the A-series, which lasted only three years. At this point, the Mack script had finally lost its tail but the basic design was still clearly evident. This was also the first time the bulldog was applied directly to the body panel. This application signified added emphasis on the symbol, since the bulldog was also carried atop the radiator shell, as before.

This scene, photographed in a yard belonging to an avid collector of Mack memorabilia in the Pacific Northwest, features the radiator shell of a B-series Mack, a line of trucks introduced in 1953. The script is embossed into a chrome bar that shows strong art deco influence in its design. The use of this style was somewhat belated, one must note, since the heyday of the era, with its far-reaching effect on both furniture and architecture, was the late thirties and the early forties.

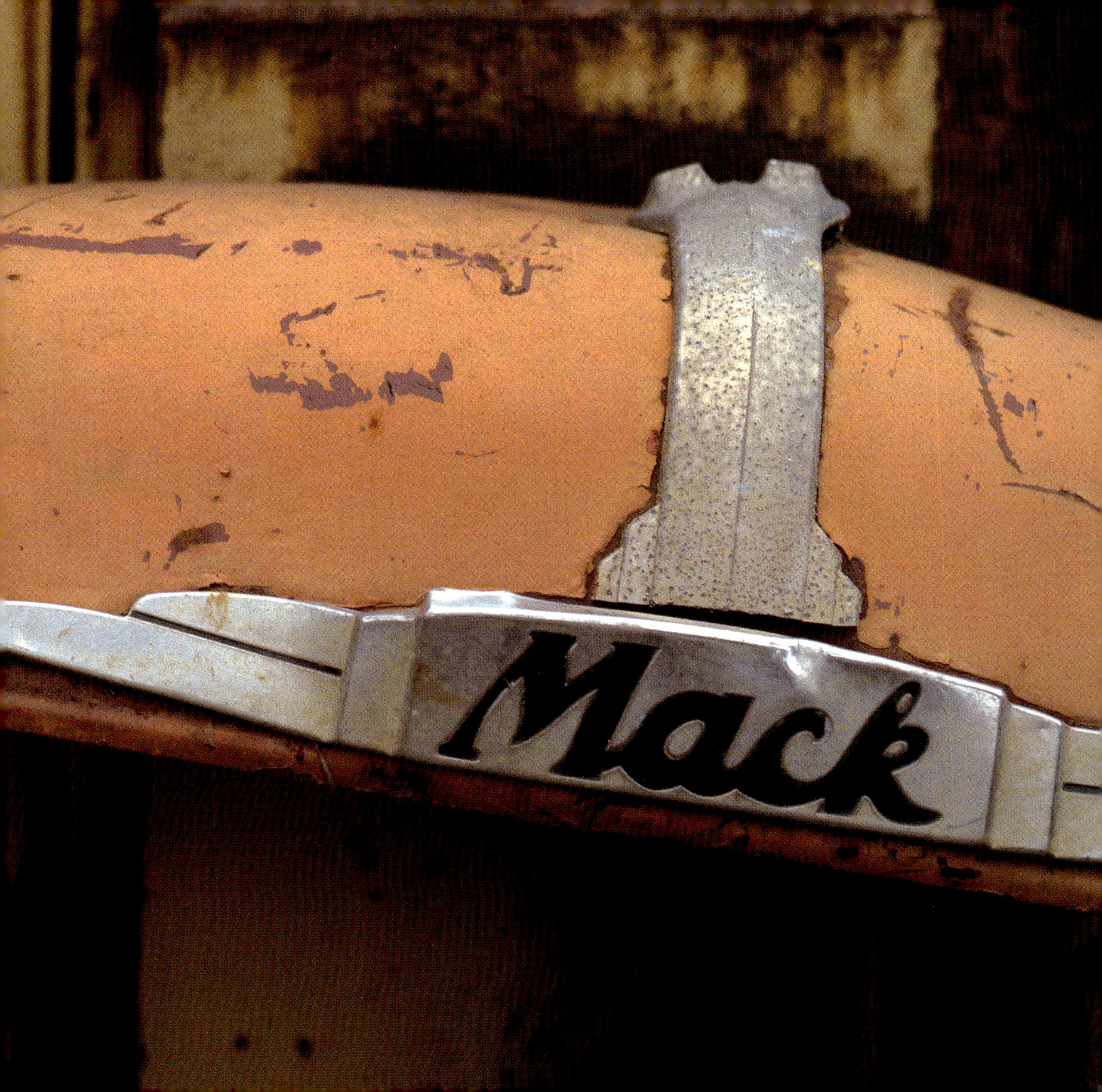
Mack

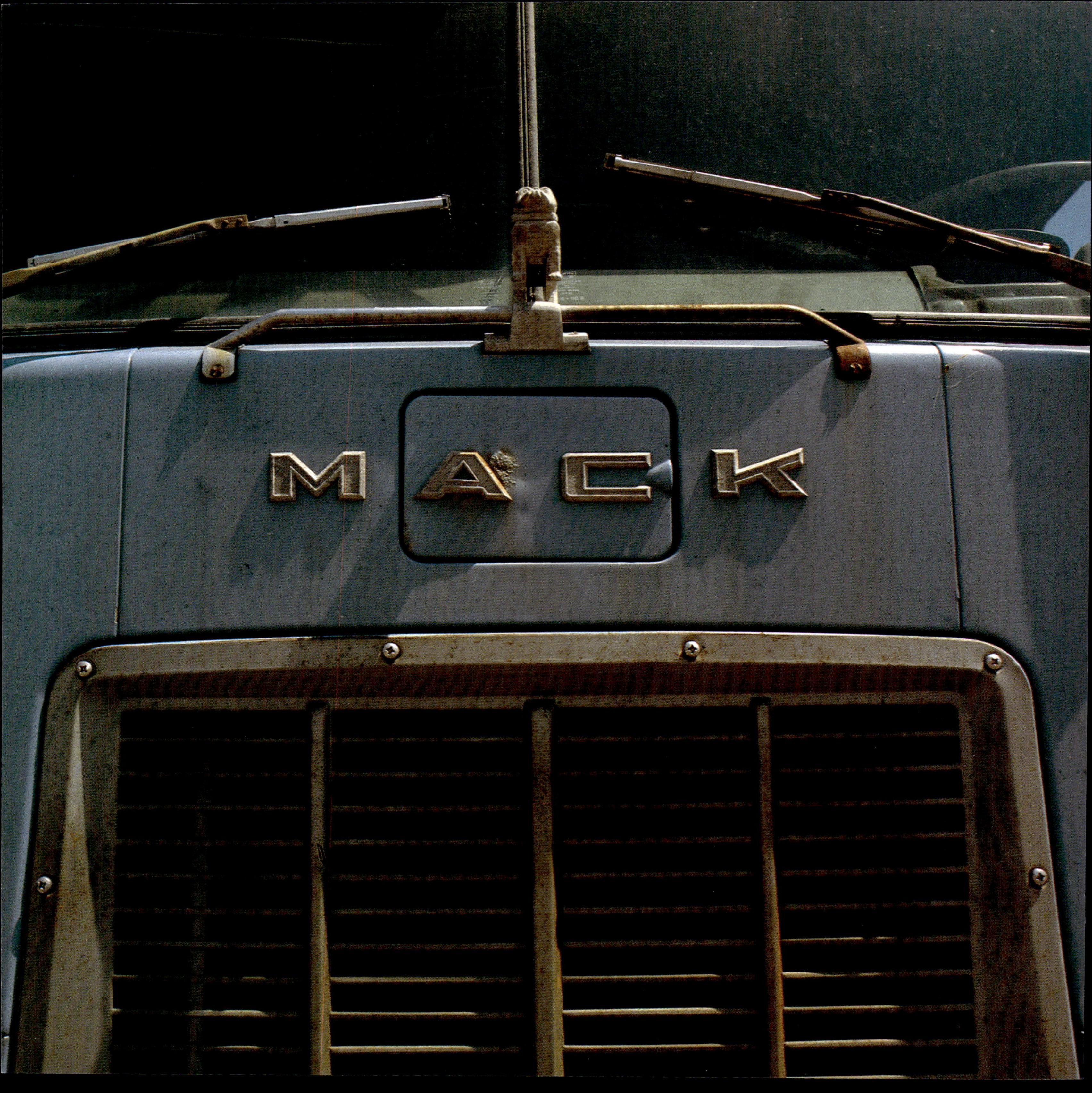
MACK

By 1962 and the introduction of the F-series cabover (cab-over-engine) trucks the traditional Mack script had disappeared. It had now been replaced by a logo made up from letters of a modern, simplistic design, as seen in the photograph to the left. The die-hard bulldog, as fierce as ever, was still on board, although, with the lack of a long hood and a conventional radiator shell, some of the traditional figurehead feeling was gone.

A new series of cabover Mack trucks was introduced in mid-year of 1975. Called the Cruise-Liner, the earliest units sported a gilded bulldog—as pictured in the photograph above—set in a beautiful cloisonne medallion. At this point the bulldog had undergone another subtle redesign. The sharp spikes on the collar had been removed. Instead the faithful old traveling companion had received a collar with a name tag. The inscription reads "Mack". What else?

Mack® TRUCKS

In the mid-sixties the formerly white coat of the Mack bulldog was changed to a brindle color. This came about as a result of an advertising campaign instigated by new, forward-looking Mack president, Zenon Hansen, who felt that the bulldog would receive even greater popular acceptance if its appearance was more life-like. At this time the Mack script was also revised slightly, as seen in the photograph to the left, picturing the design as it appears on the mud guard of a Mack truck, road dirt and all.

The R-series Macks, introduced in 1965, boasted Mack badges of two different designs; the modern logo, introduced a few years earlier on the F-series, was carried above the radiator opening; the newly revised Mack script, reminiscent of the traditional logo of the early days, was carried on the side of the hood, as seen in the photograph above. Mack truckers showed their patriotic spirit during the 1976 Bicentennial Celebration by displaying this Stars and Stripes decal—a collector's item today.

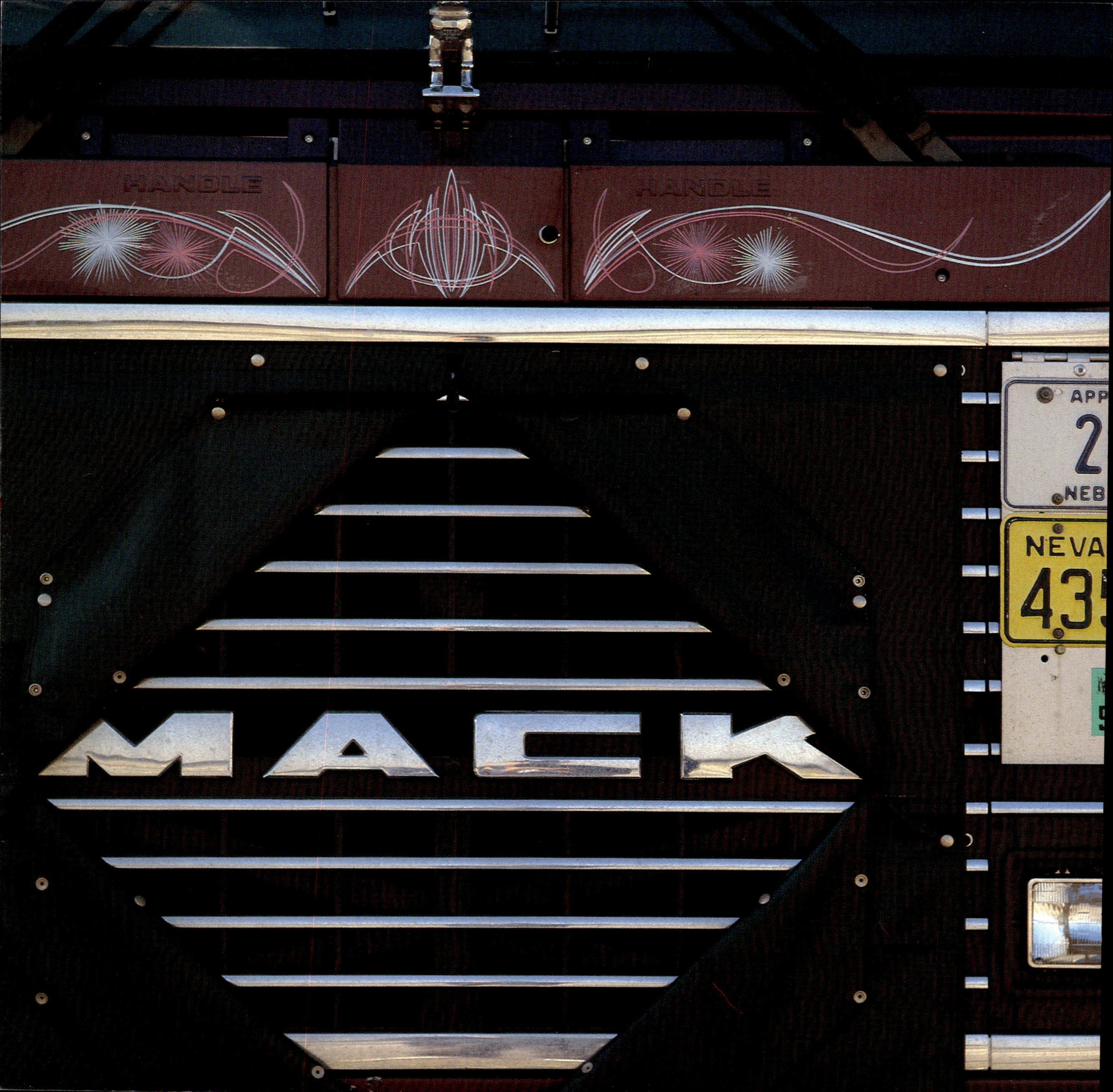
HANDLE
HANDLE
MACK

The 1982 introduction of Mack's new Ultra-Liner brought further changes to the Mack logo. It has now reached a level of ultimate simplicity and is perfectly worthy of the true marvel of the modern truck it decorates. However sparse the face of today's high-tech fashion, or perhaps because of it, the trucker enjoys adding his own touches, as in the picture on this spread, with its elaborate expression of pinstripe happiness. Add a variety of license plates, top it off with a winter front, and the picture is complete.

Next page
The Mack designers are not beyond expressing some graphic exuberance of their own, as evidenced by the photograph on the following spread. Featured here is a close-up of the limited edition Mack Magnum, which is a further refinement of Mack's already powerful and advanced Super-Liner. Contrasting the awesomeness of the pitch-black cabin and hood with the hot and fiery intensity of red and orange—all tied together by a name that rings tough with macho muscle—the Magnum is destined to become a hot item in more than one sense.

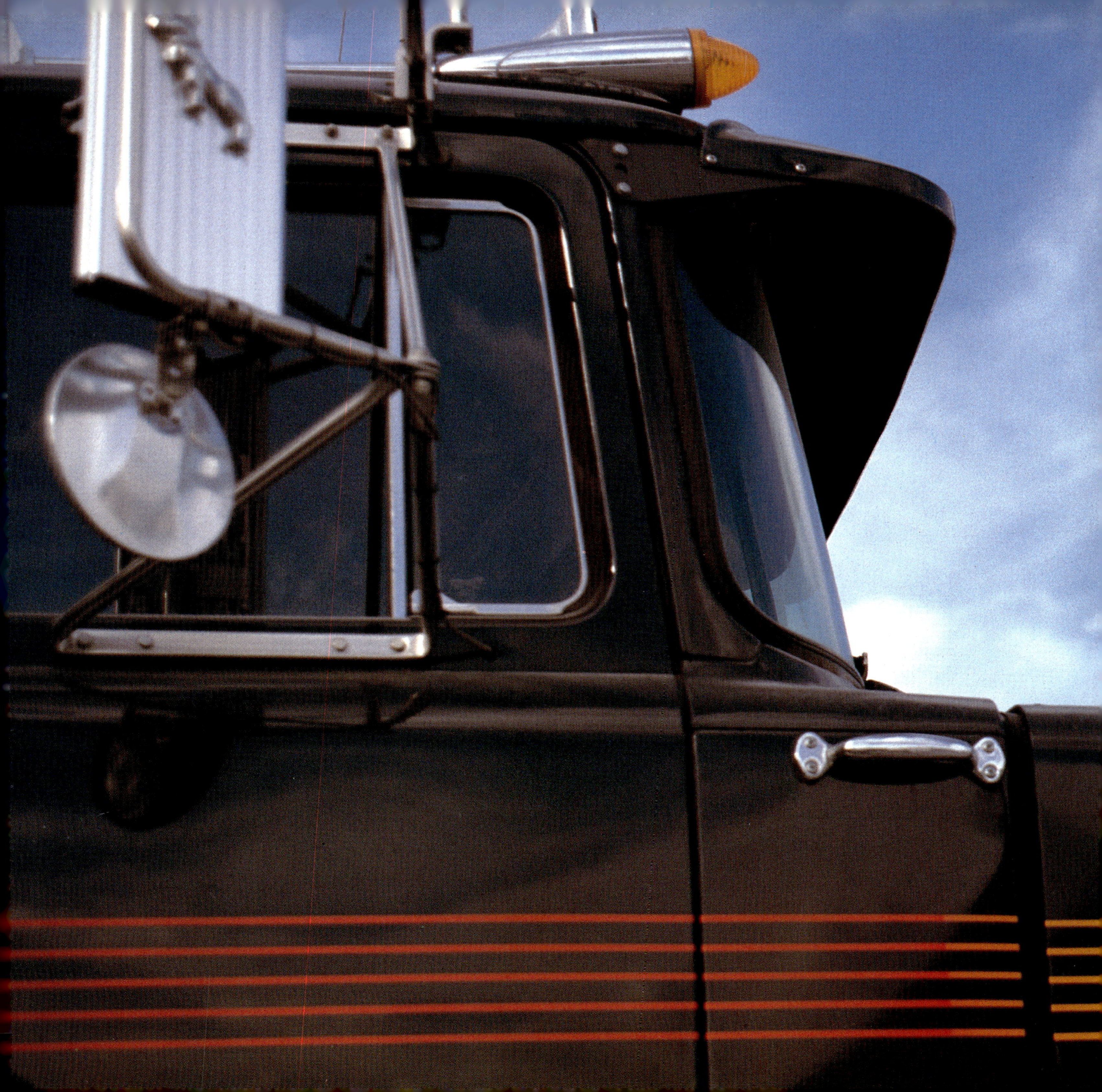

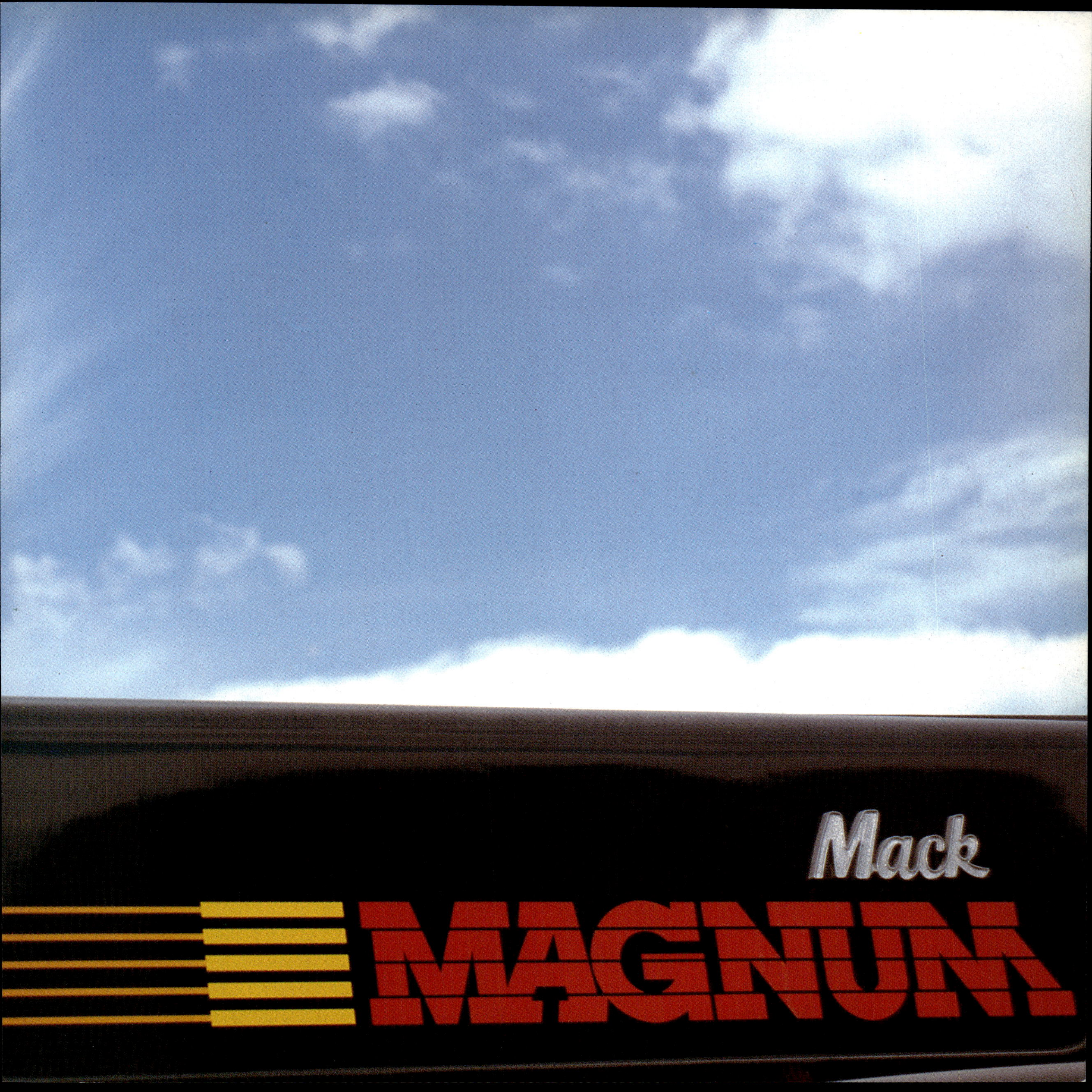
Mack
MAGNUM

Family Mack

When Mack runs in the blood

Bob Brown received an early introduction to Macks and trucking; not only did his father drive Macks for the Highway Department in the state of Washington, but his name was Mack. And it was not just a nickname, he was actually christened Mack. No wonder then that Bob's life today is centered around trucks and trucking, and Mack trucks in particular. He now owns a wrecking yard located on Pacific Highway 99, right on the line between Washington's King and Pierce Counties. Out of this location he operates a business that deals in used truck parts, with Macks being his specialty. He enjoys what he is doing for a living. But what he enjoys the most is collecting and restoring old Macks. His yard shows ample evidence of this avocation, with old wheels and frames and engines occupying every square foot of space, and of course entire trucks, both restored and shiny as well as seemingly forgotten and rusty. It is an intriguing thing this—having Mack running in one's blood.

Bob Brown is a busy man, and a man whose thoughts are not much occupied with vanity; it is not easy to pin him down for a portrait. Combining the photo session with a moment's rest on the running board of one of his beloved Macks makes it seem somewhat less of a chore to him. This Mack is his workhorse, a 1948 model HQ, mainly used for hauling scrap.

WN
PT., INC.

WASH.
7-2566

Mack

Pictured here is an example of the treasures unearthed by Bob Brown. He located this 1928 model AB in Shelton, Washington. Originally used for logging, it had been left to waste in the forest and Brown was forced to cut down a number of trees to get it out. The AB, as far as numbers are concerned, was the most successful of Mack's post-World War I trucks. Nicknamed the "Baby Mack" and introduced in 1914, 52,000 units had been built when production ended in 1936. In 1923 the AB received a new radiator. Extending a couple of inches above the hood, it lent a distinctive new look to the small Mack.

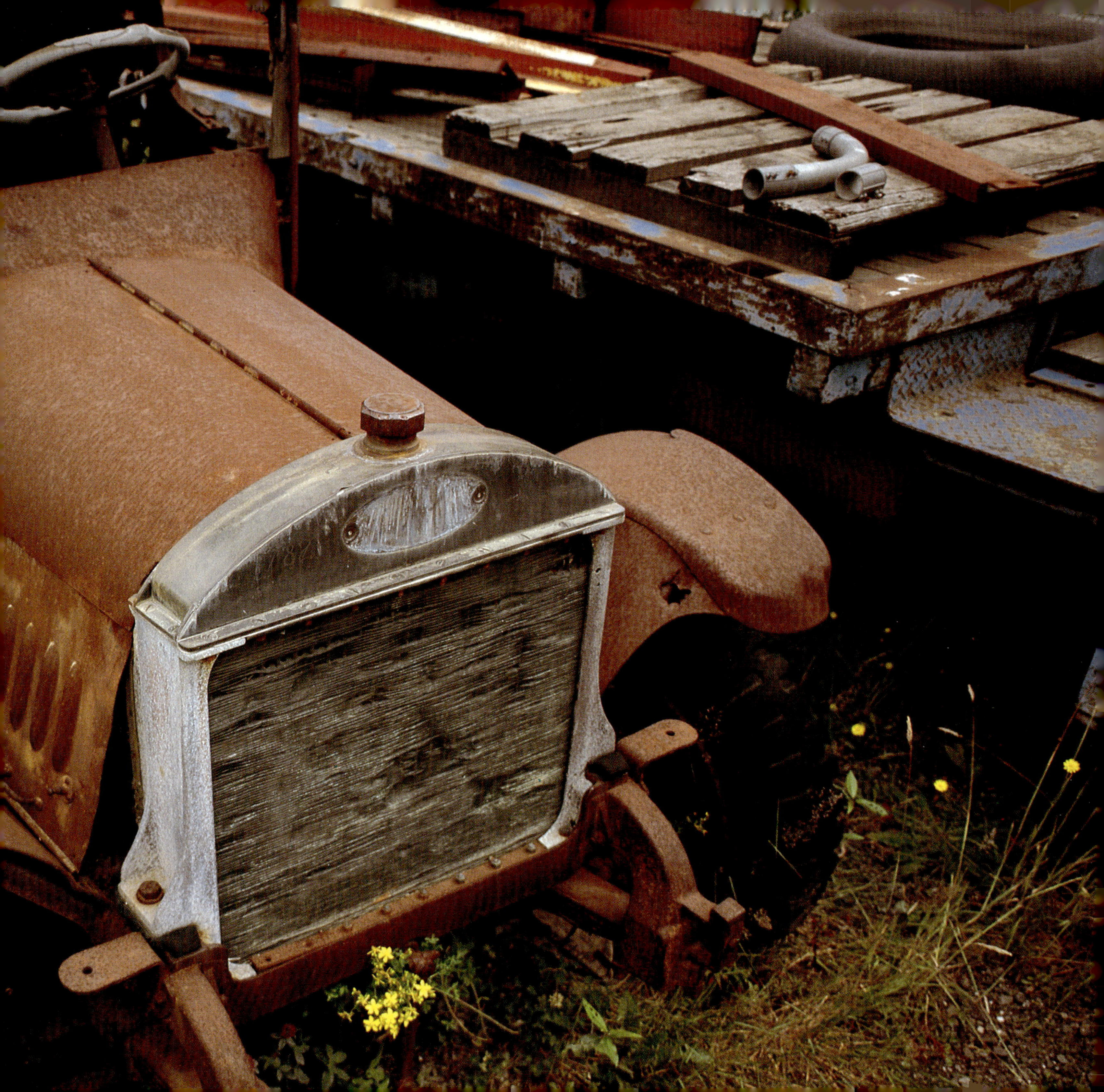

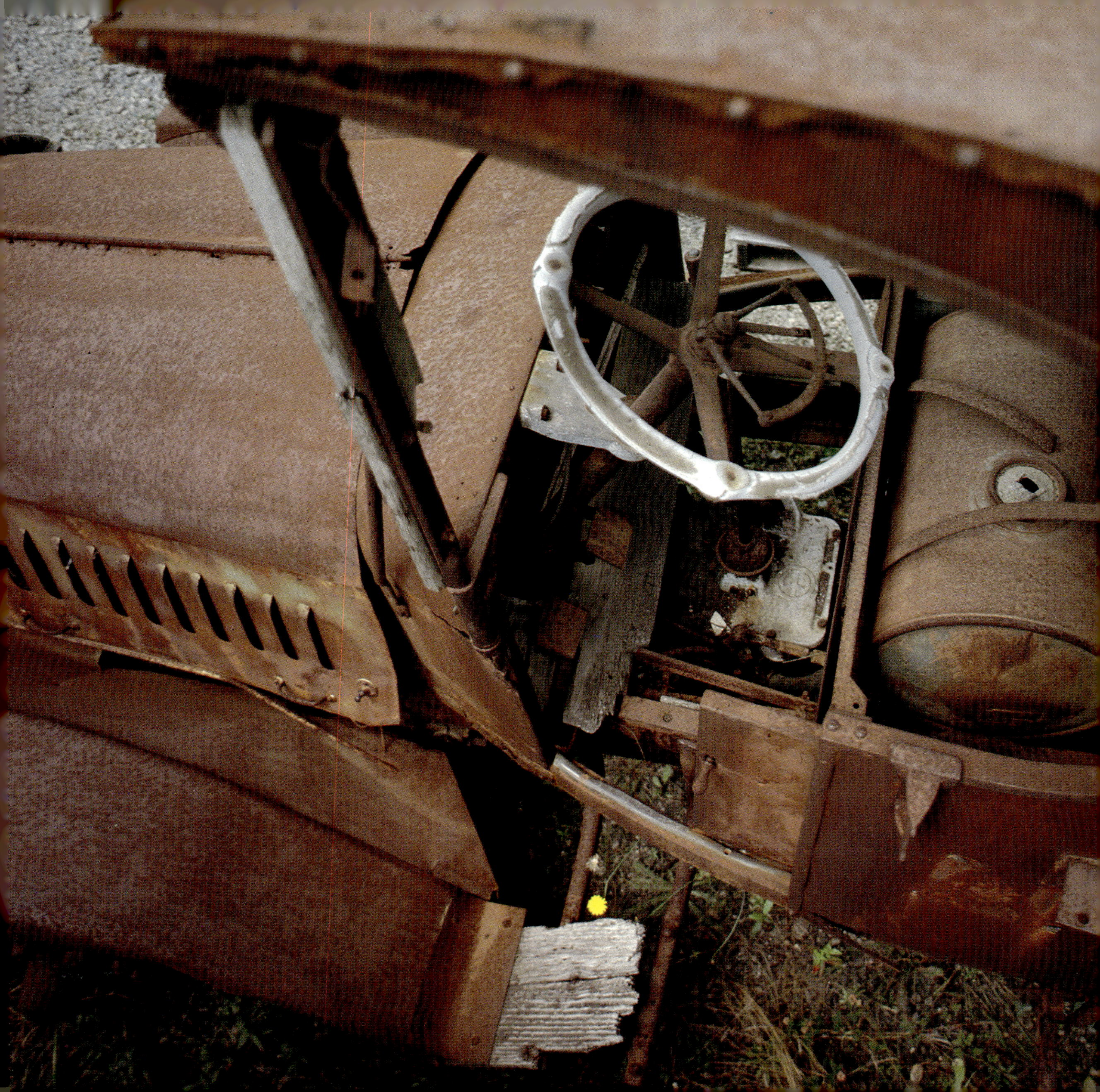

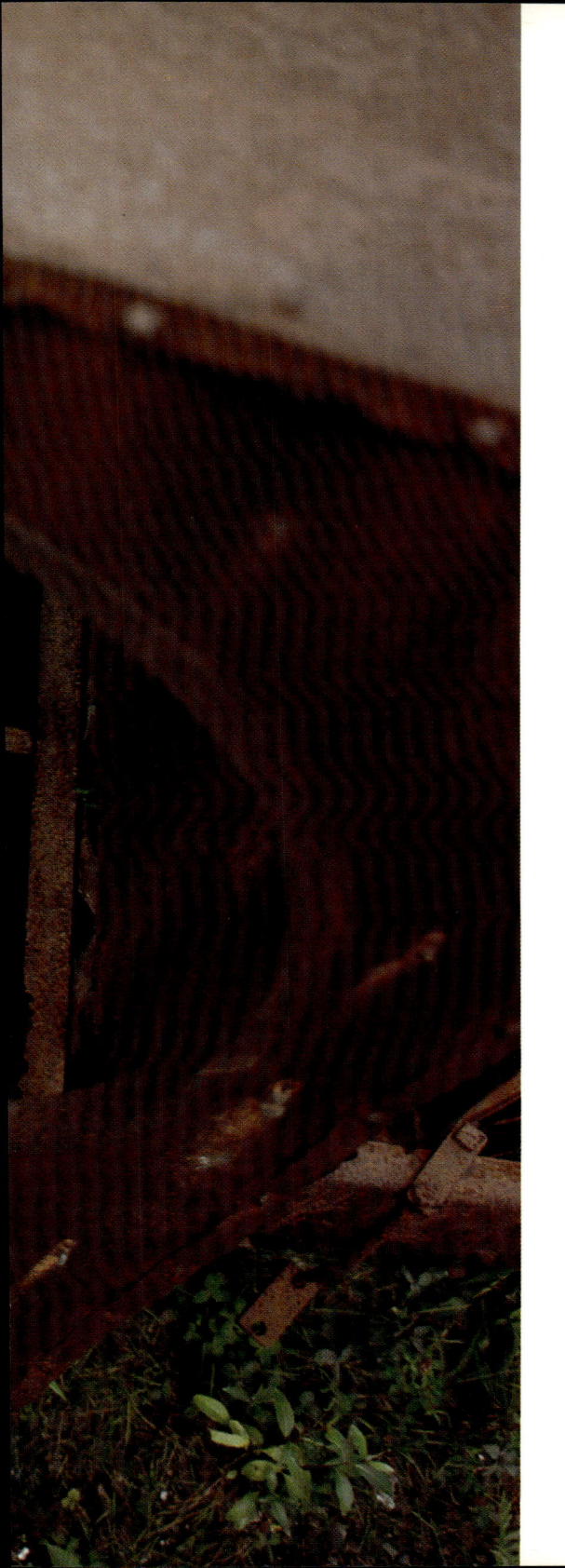

The photograph on this spread shows Bob Brown's 1928 AB from another angle. Note that the fuel tank was located under the driver's seat, which had to be removed when gasoline was taken on. The 3-speed transmission, as well as the 4-cylinder engine, were both built by Mack. Early trucks were rated at about 30 bhp, which gave the Baby Mack a working speed of about 16 mph. By 1928 the engine had received new cylinder heads, referred to as "hard hats". Power had now risen to about 40 bhp, and working speed to about 25 mph.

Next page
Bob Brown does not collect just unrestored trucks; the photograph on the following spread shows one of his superbly restored examples, a survivor with only 48,000 miles on the odometer. Brown found the Mack veteran in Carnation, Washington, where it served as a water truck. The show stopper is a 1936 model BM, a type first introduced in 1932. At that time a BC engine was used, but later the 400 ci, 6-cylinder CE unit was fitted, which meant that as much as 108 hp could be transmitted through the 5-speed tranny.

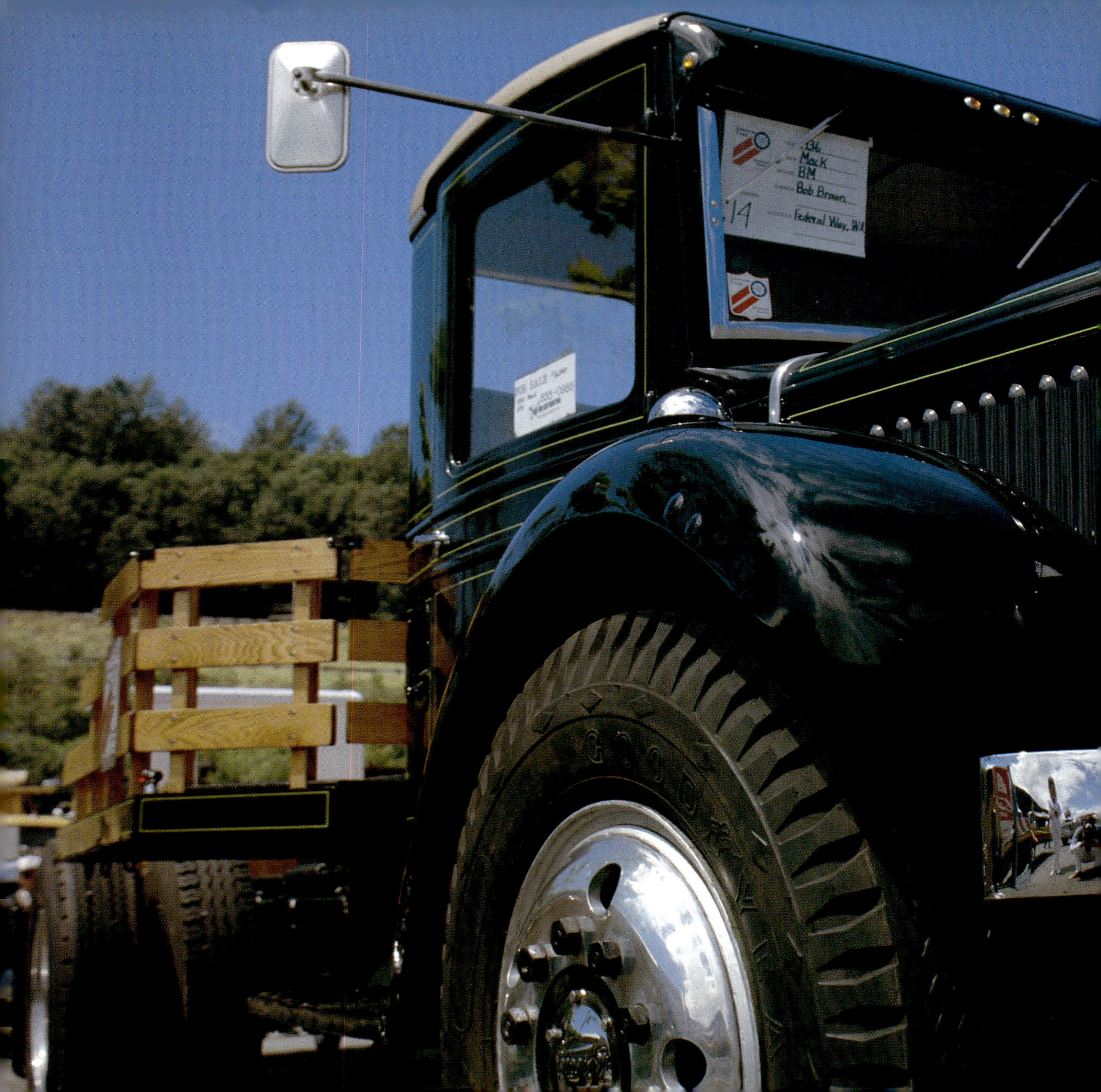

36
Mack
BM
Bob Brown
14
Federal Way, WA

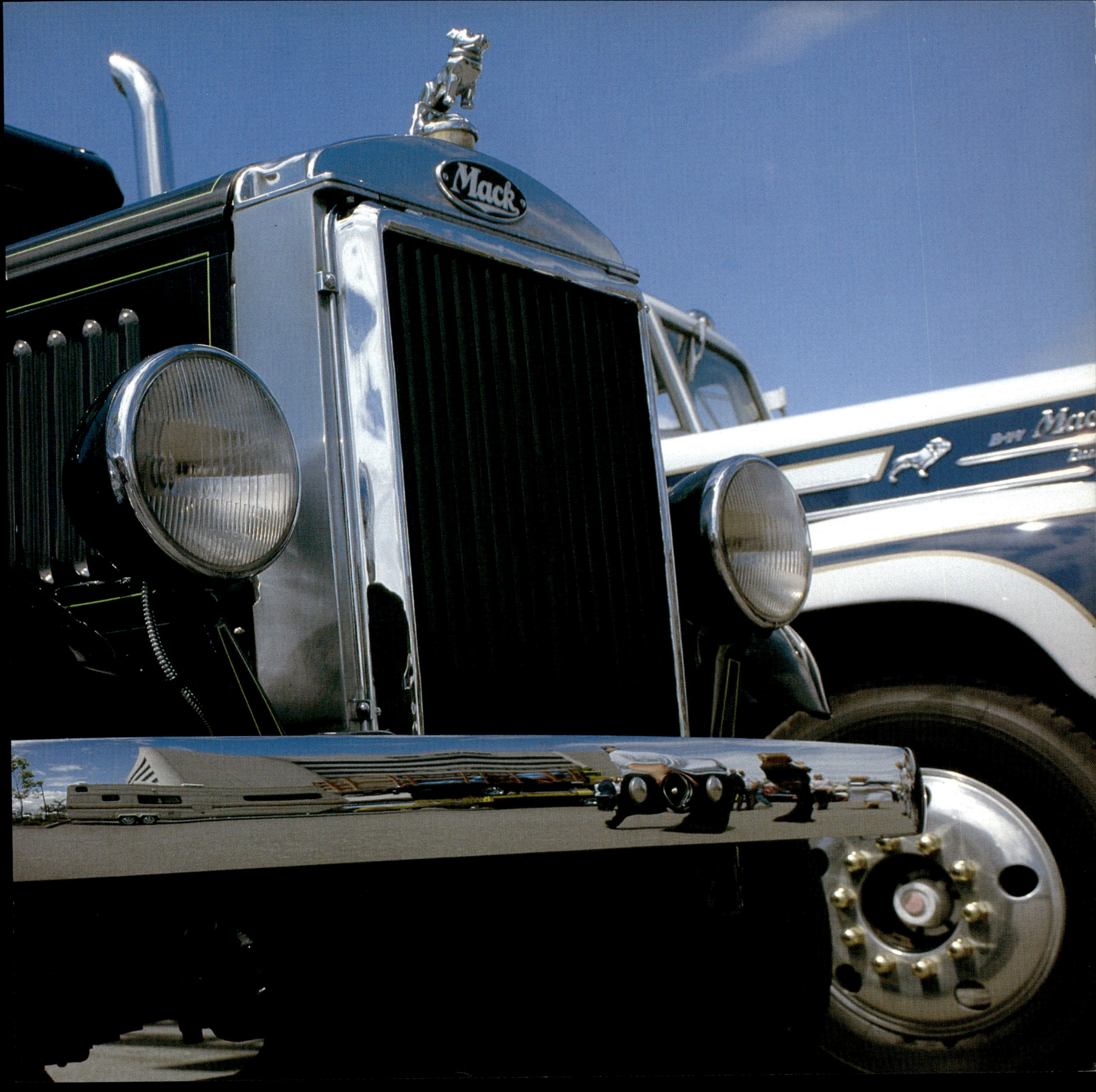
Mack

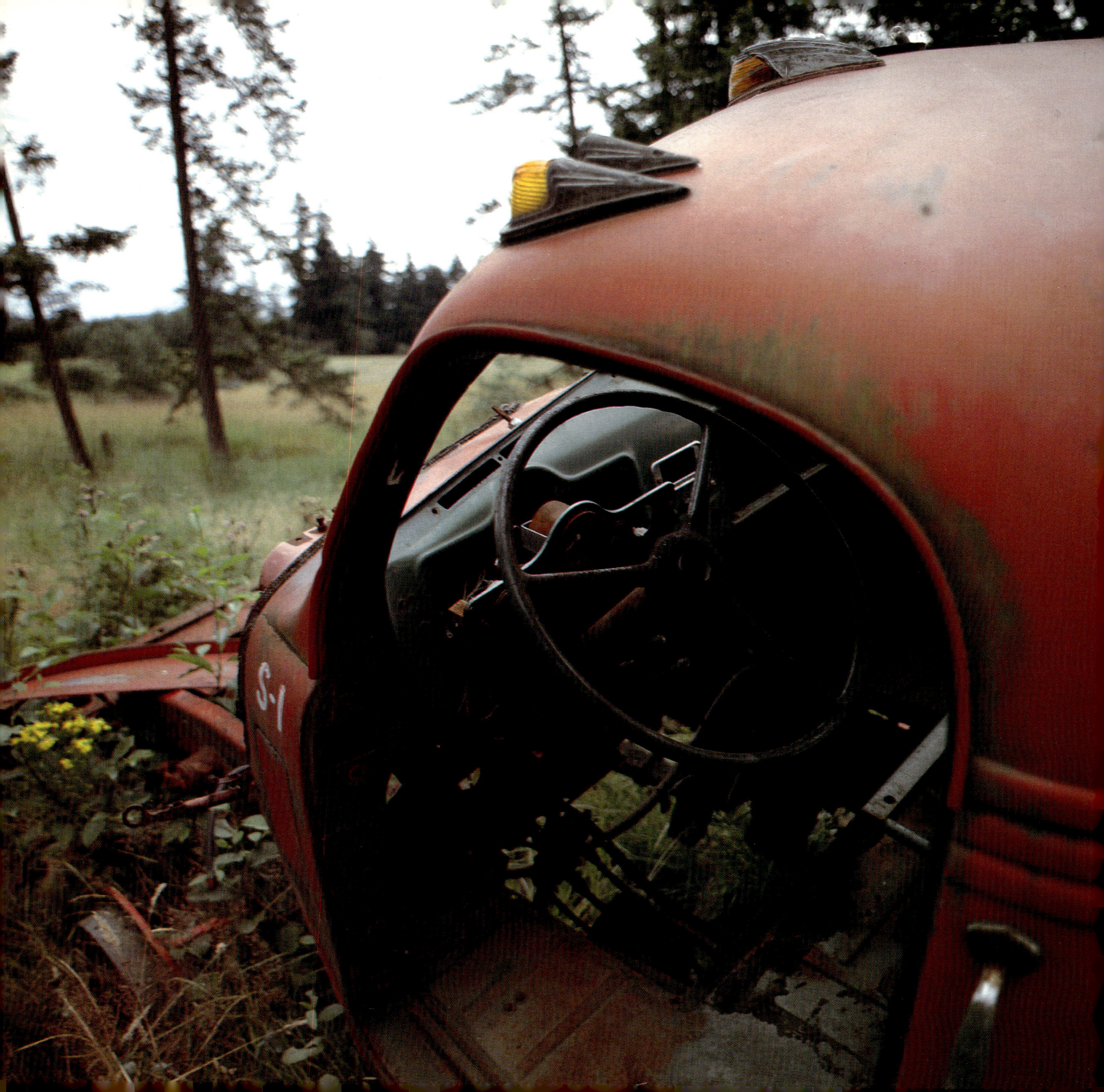
S-1

At the edges of Bob Brown's intriguing Washington spread, Mother Nature and Mack Memorabilia have formed alliances that manage to charm even the hardiest of nostalgia buffs. In the photograph to the left the abandoned cab of a 1954 model B-61 has been left to rest in peace. However, every year a few more of the salvageable goodies are taken off and put to good use, thus prolonging the life of others.

Another area of Brown's salvage yard yields the fascinating collection of unique old Mack engines featured in the photograph above. Lounging in the foreground, a 231 ci gasoline engine from 1946, complete with 10-speed transmission and all. Seen next, a 510 ci engine from 1950, also used to consuming gasoline. In the background, an example of an early Mack diesel engine, a 672 ci unit from 1948. Note the beautifully finned intake manifold.

47

Fifties' Mack

Bold faces of a boom era

After the stop-and-go growth of the twenties, thirties and forties, with their periods of depression and war, the fifties experienced an expansion of the trucking industry of never-before-seen proportions. The settling of the rules and regulations that govern that industry, as well as the completion of the interstate highway system, were important factors behind this phenomenal growth. And Mack was as always at the forefront, meeting the demand for new machinery with a steady flow of new models. The particular line of trucks that carried the Mack flag forward during this period was the B-series. Introduced in 1953, this new line of trucks with its massive, rounded hoods and fenders, spelled Mack to the masses the same way the venerable AC Bulldog had barked Mack during the between-the-wars era. During the production period, which for some models lasted until 1966, almost 150,000 units were manufactured, representing an array of nearly 70 variations on the theme. The fifties was indeed a boom era—to trucking in general, and to Mack in particular.

Shown in this photograph is an example of the most numerous of all models in the B-series, the B-61, of which nearly 50,000 units were built. This particular workhorse is from 1954 and features an exhaust-turbocharged 673 ci Thermodyne, producing 205 hp. There are two gearboxes on board, one featuring five speeds, another featuring three speeds. All this gives the well-equipped operator a menu of 15 gears to choose from.

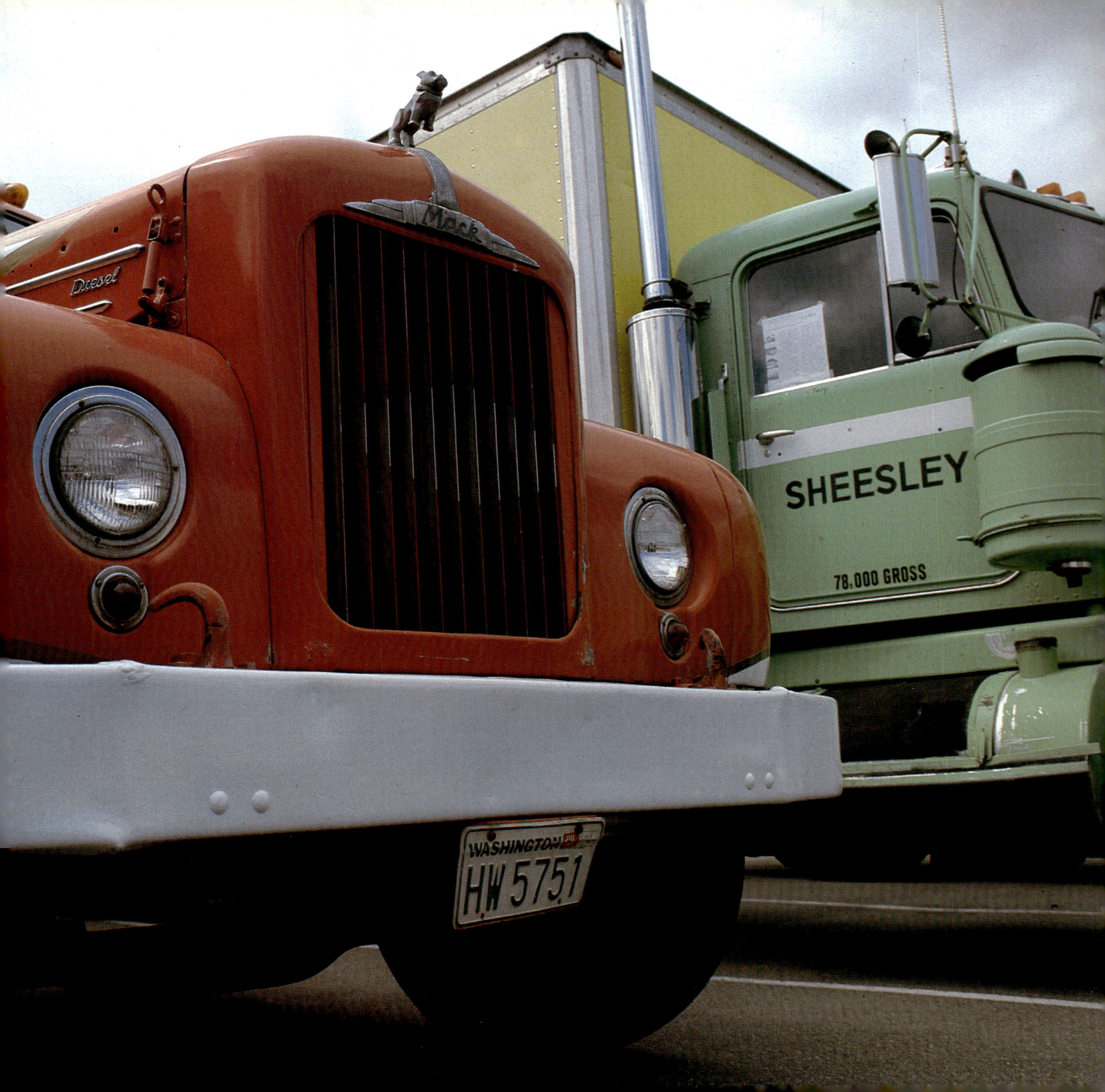
Mack
Diesel
SHEESLEY
78,000 GROSS
WASHINGTON
HW 5751

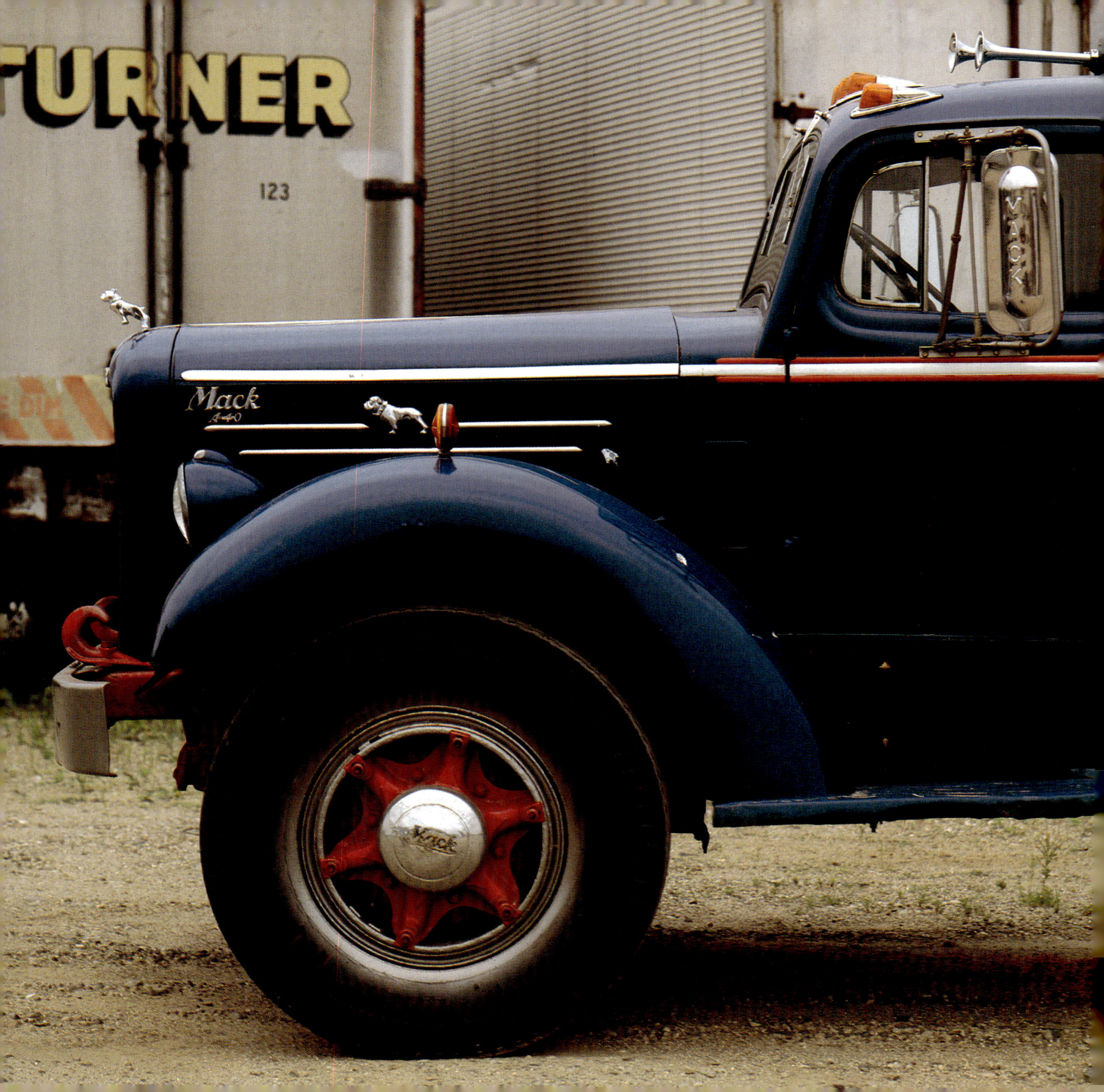
TURNER
123
Mack
A40
MACK
Mack

The A-series was introduced in 1950 and carried forward the look of the models produced by Mack immediately before and after the Second World War. The example pictured on this spread is an A-40 from 1952. Hidden under the hood is Mack's 405 ci flathead Magnadyne gasoline engine. The beautiful survivor is still in virtually original shape—not counting a recent paint job—and was for many years the dependable workhorse of a construction firm in Forest Lake, Minnesota. Production of the A-40 ended in 1953, after a total of 7,666 units had been placed on the road.

Mack, with its deep roots in pioneer engineering, has always taken the art of building trucks very seriously. Today the company is one of the few in the industry still manufacturing its own engines, transmissions and suspensions. The picture above, featuring a portion of the famous Mack Track Camelback suspension, as mounted on the 1952 A-40 pictured to the left, illustrates the fact that functional design is not only functional, but beautiful. Form and function united.

Next page
Another beautiful survivor, a low-volume B-70, decorates the following spread. The B-series was engineered with a wider front, which resulted in better access to the machinery and allowed bigger engines to be fitted. The pictured example is powered by Mack's largest gasoline unit with a capacity of 707 ci and an output of close to 200 bhp. In the mid-fifties, with lingering prejudice towards the diesel, many truckers preferred the quicker response of gasoline power. The B-70 is a rare and desirable machine indeed. During its term of manufacture, 1953 to 1966, only 1,073 units were built.

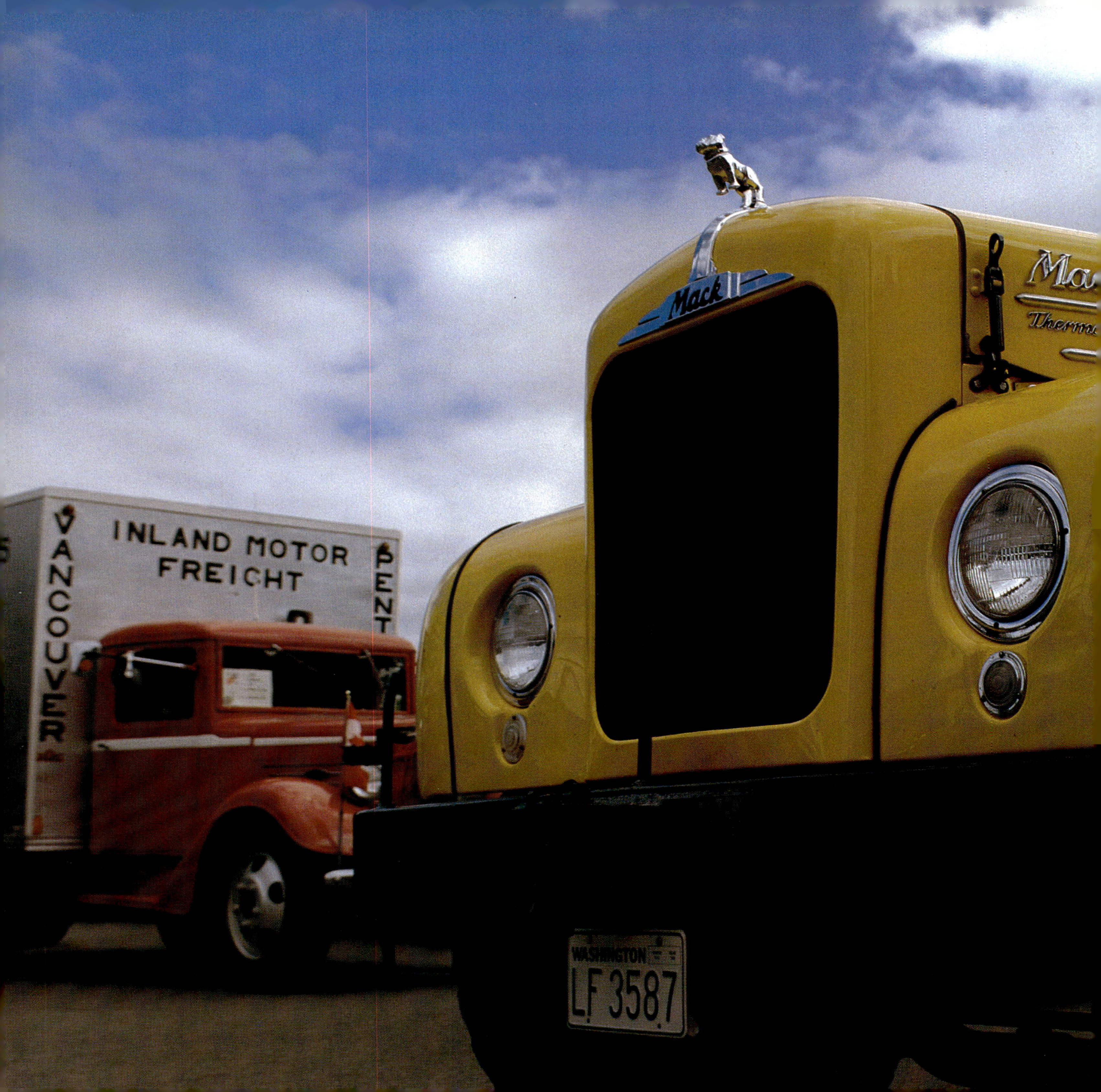
INLAND MOTOR
FREIGHT
VANCOUVER
Mack
WASHINGTON
LF 3587

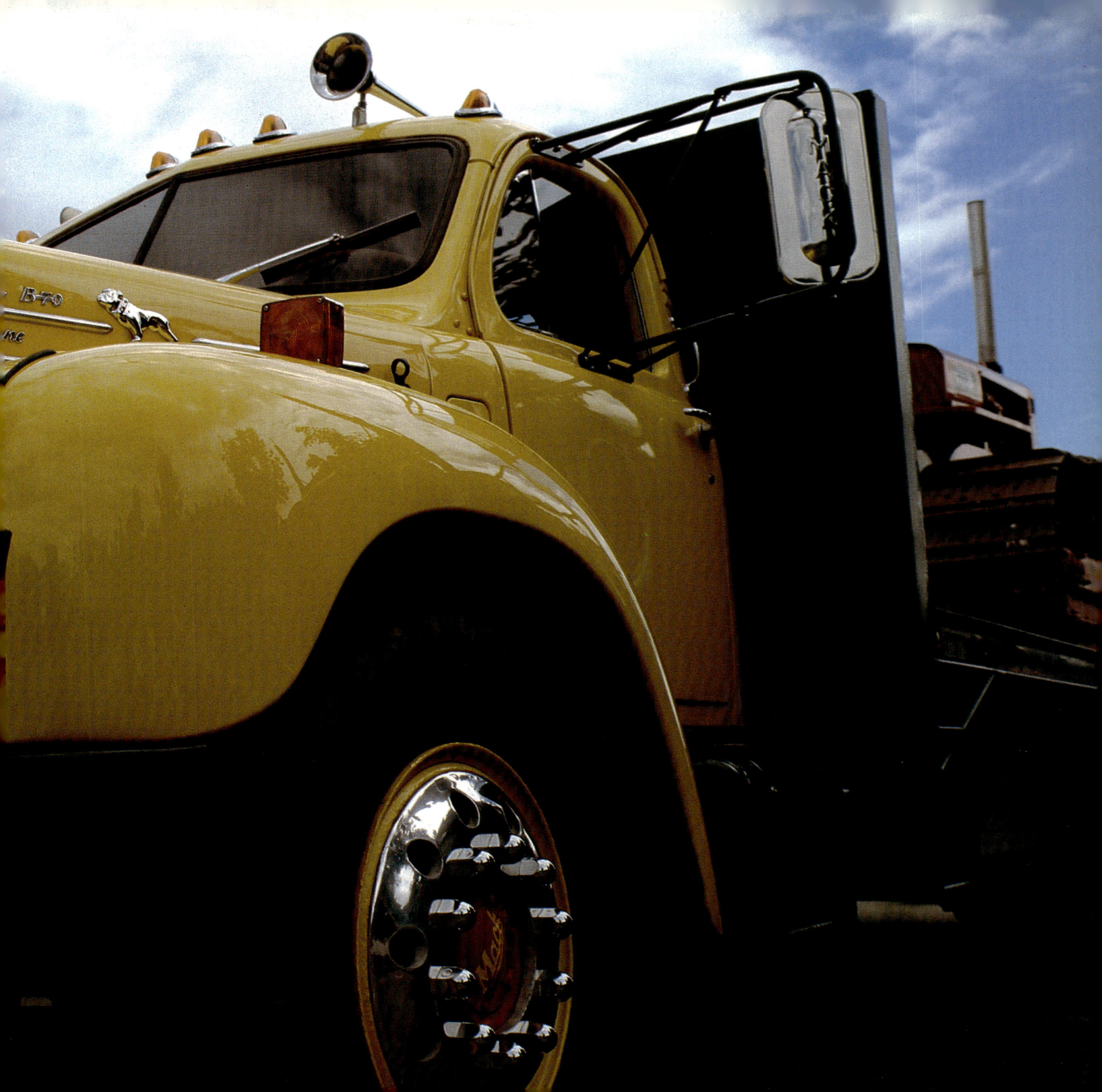
B-70
MACK

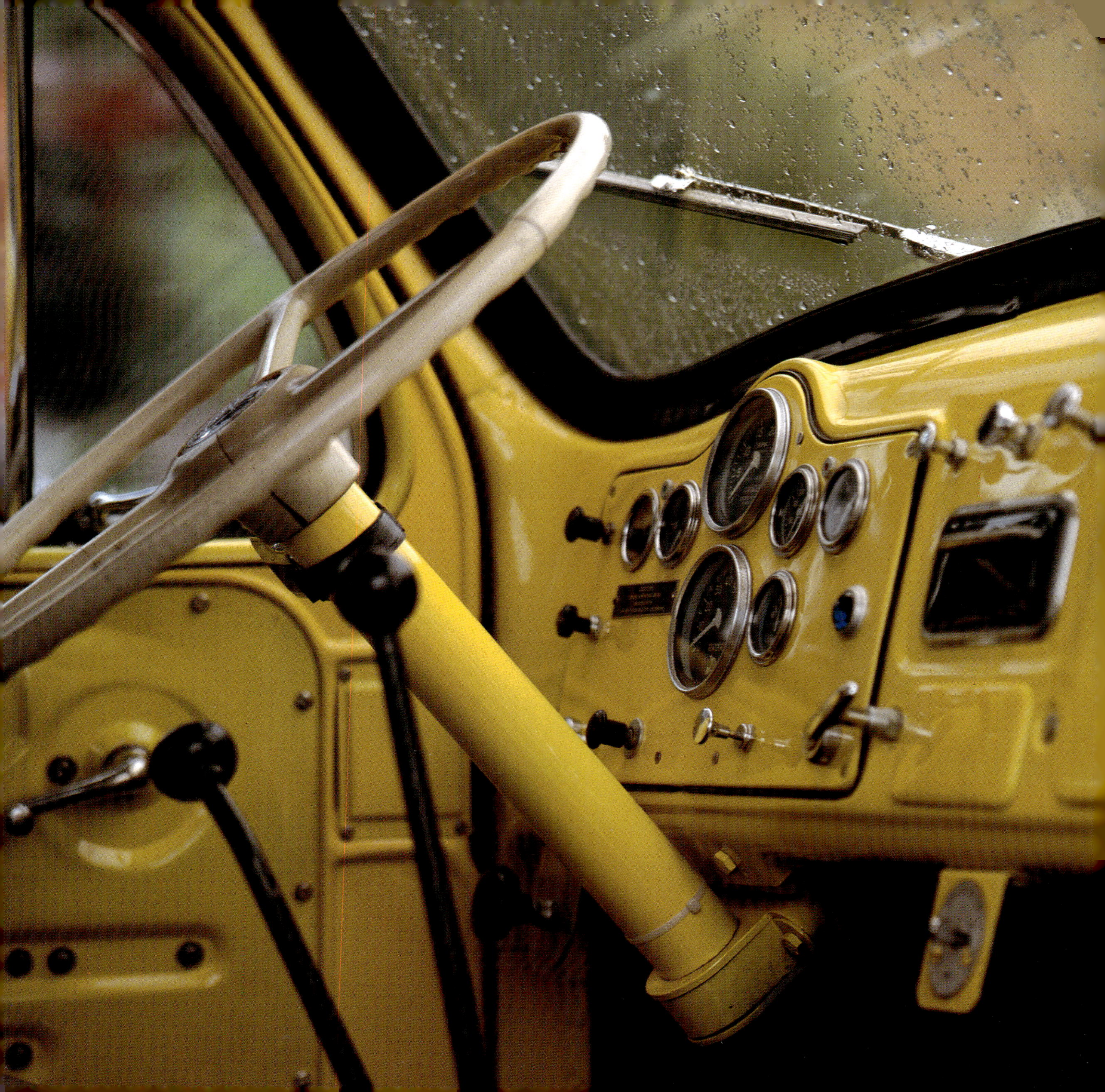

The yellow paint scheme chosen by the restorer of this B-70, gives the interior of the cab a cheerful ambiance, as seen in the photograph to the left. In the B-series the driver's work place had been the object of much study. The efforts resulted in, among other improvements, better ventilation. A further result of the Mack designer's attention to detail was the V-type windshield, which was set at a particular angle, said to cut down on dangerous glare. Furthermore, the instruments had been re-arranged on a cluster-panel, for easy access.

While six-spoked wheels (of the type seen on page 50) were sufficiently strong for the type of trucking taking place in the East, truckers in the West tended to favor the Budd-type wheels, as seen in the photograph above. These wheels, with their ten sturdy nuts, as opposed to the six used on the former, were considerably stronger. In this picture the aluminum has been polished to a highly decorative sheen. The cover of the wheel bearing still carries the Mack logo from the old days.

55

Diesel

Jake C. DeWitt
Tolleson, Ariz.

Featured on this spread is one of the most legendary of the fifties' Macks, the awe-inspiring LTL. Introduced in 1947 the model was especially designed to meet the needs of the Western states. In contrast to the laws governing trucking in the East, where the vehicles had to conform to restrictions on length, the requirements of the West focused on weight rather than length. This was the reason for the lightweight metals used in cabs and frames, the long wheelbases, and the long hoods. These hid the most impressive powerhouses in the industry, such as the 300 hp supercharged Cummins. The example shown here is still in active duty, although highly modified to allow efficient operation. Production ended in 1955, after just over 2,000 units had been built.

Mack Woman

She is the "Lumber Gal"

Things connected with trucks and trucking have always been one of the prime areas of male dominance, with the world of the trucker centering around concepts like horsepower, heavy-duty, and muscle. But this never deterred Mary Hitchings. Right from the beginning, already while she grew up on her parents' farm in Santa Rosa, California, she preferred trucks over dolls. Later she married a man who knew how to drive trucks. But perhaps he did not know it well enough, for, as husband Andrew relates the story, she always told him how to do it, until one day he got tired of the unsolicited advice and asked her to get a truck of her own. She did. And she did more than that. She got lots of them. In all different makes and colors. After sampling the lot, she decided on the one with the bulldog on the radiator. Today her collection reflects that preference; there are as many as 36 Macks sprawling on the Hitchings spread. Mary is living proof that you do not have to be a man to drive a truck—behind the wheel, on the CB radio, she is the "Lumber Gal".

Here is Mary Hitchings boarding her beautiful 1952 Mack LJ. The truck classic has been serving her faithfully for the past two decades. In the old days, for a number of years—and six days a week—she drove it back and forth to Fort Bragg, where she took on a load of the merchandise the Hitchings sold in their lumber yard. Nowadays, since the lumber companies deliver themselves, using their own trucks, the LJ has been semi-retired, and is most of the time driven just for the pleasure of it.

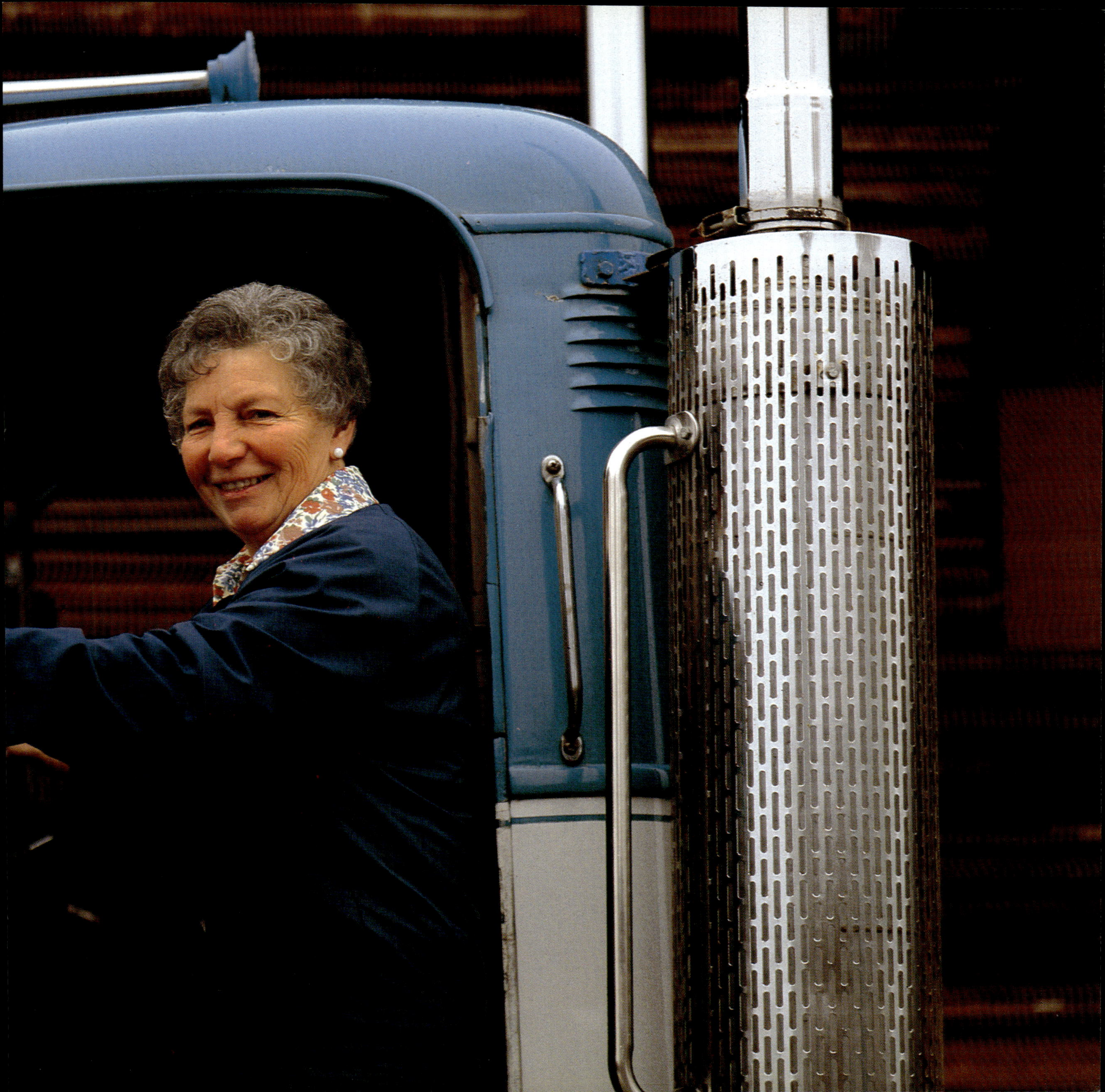

Next to the EH, the LJ was the most common of the Mack models built during the years just before the war, and the decade following immediately afterward. The LJ was actually brought out in 1940; when production ended in 1956, a total of 13,931 units had been manufactured. In this photograph Mary Hitchings guides her well-preserved treasure through the outskirts of Santa Rosa—located just north of San Francisco—while the neighborhood is still sleepy with morning fog. These trucks did not have power steering. "No problem," Mary says. "I have power steering in my arms," she adds with a smile.

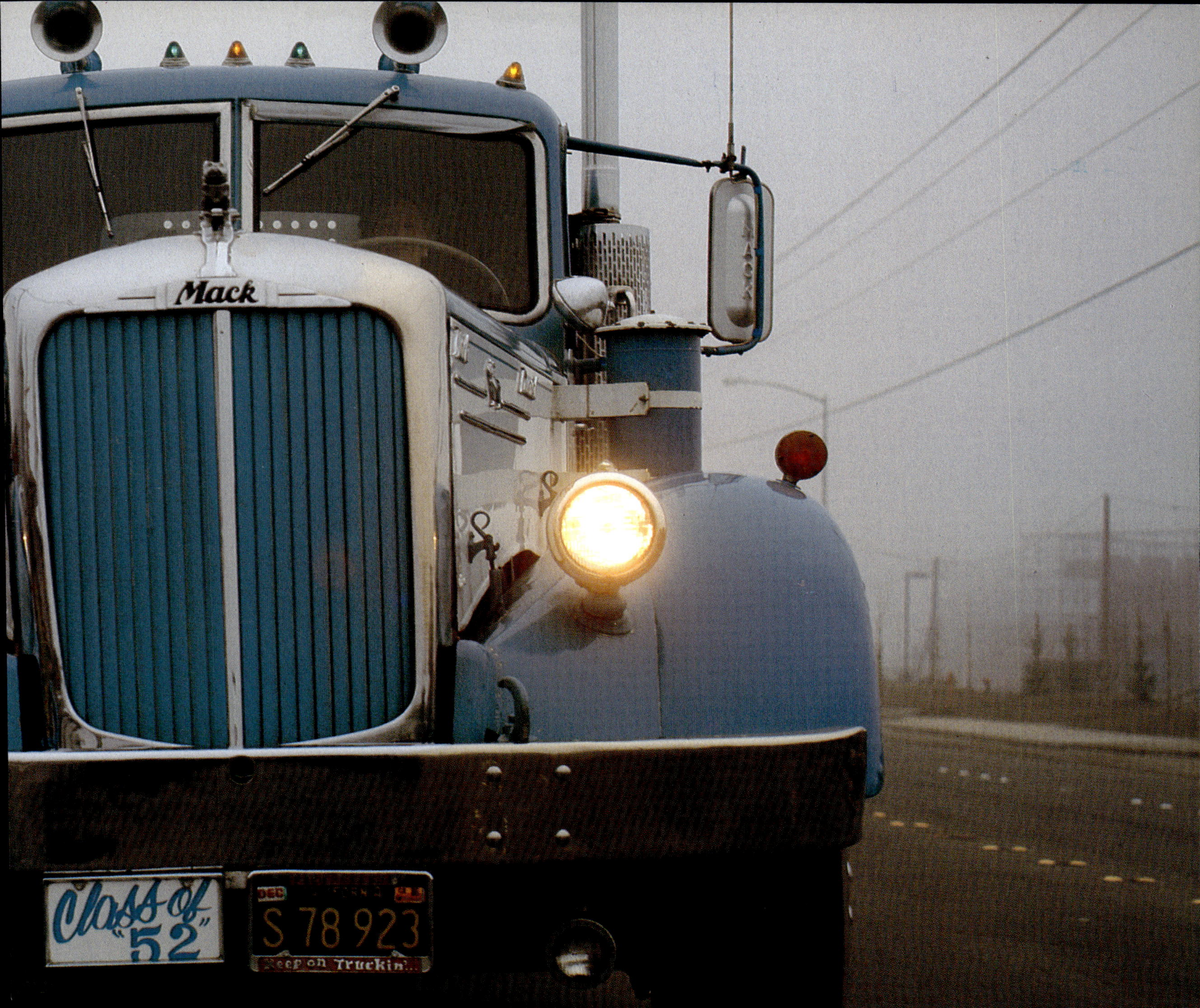
Mack
Class of "52"
S 78 923
Keep on Truckin'

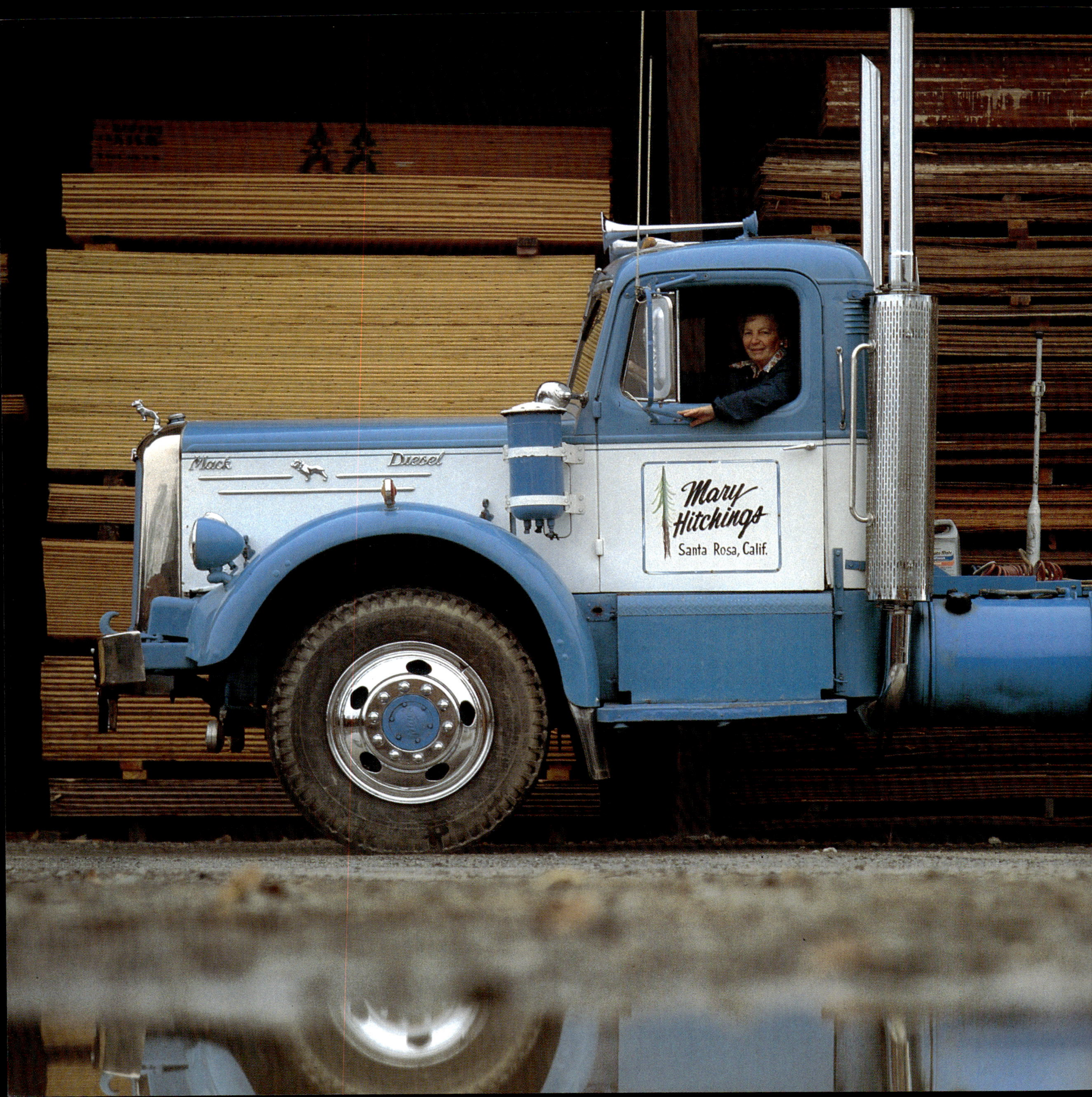

Mack
Diesel
Mary Hitchings
Santa Rosa, Calif.

Painted in Mary Hitchings' favorite colors—blue and white—and equipped with dual exhaust stacks, twin mirrors, and twin horns and spotlights, this beautiful old Mack tractor, which has more than one million miles on the odometer, still cuts a most impressive profile. The long hood hides a huge Cummins diesel with a displacement of 855 ci and an output of 270 hp. All this power is transmitted to the road through a 5-speed main and a 3-speed auxiliary gearbox. A handful for any man—or woman.

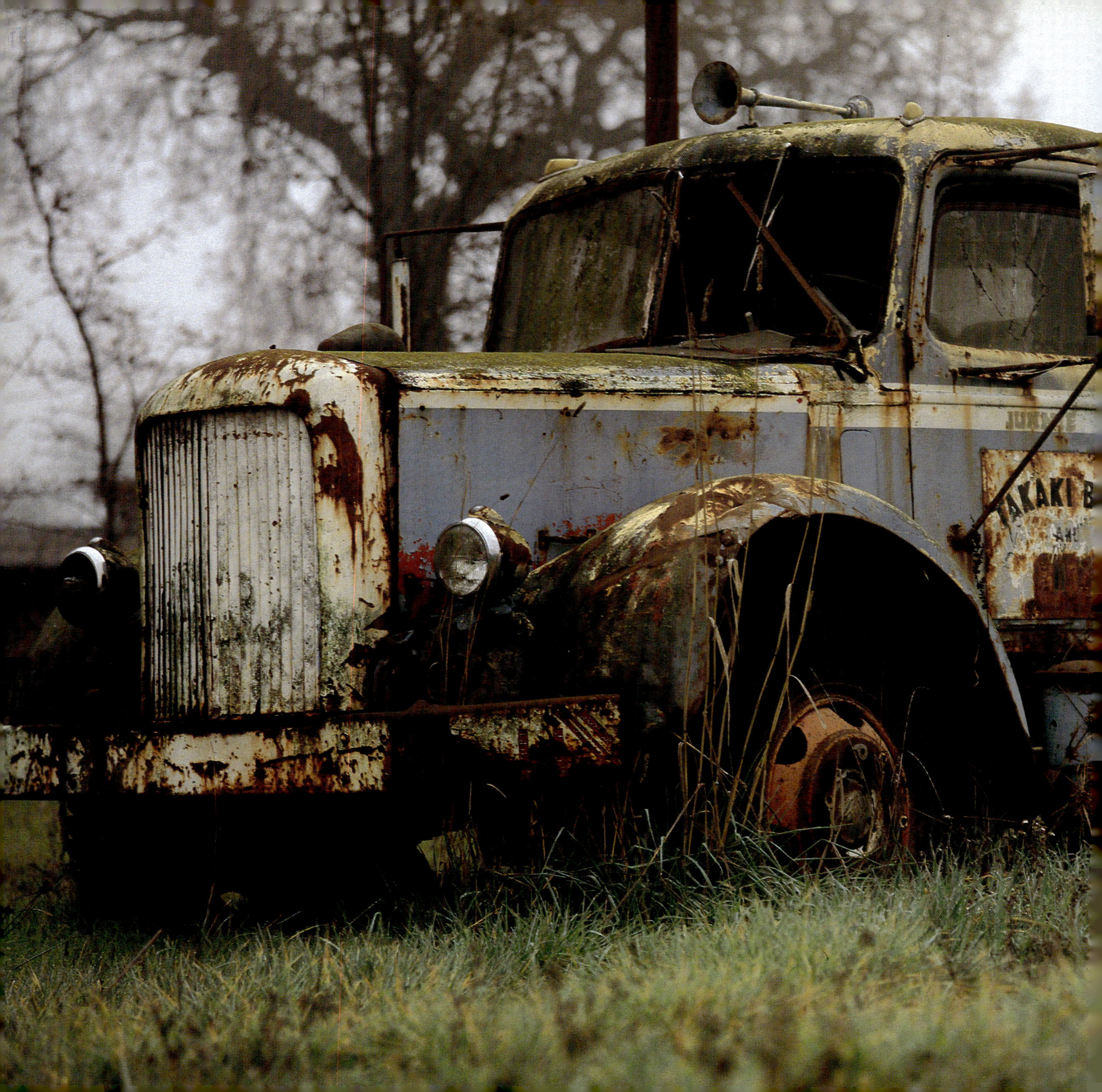
TAKAKI B

Pictured in the photograph on this page is Mary Hitchings' first Mack, a 1941 LH. The mossy old survivor is quite a rarity; production, which began in 1940 and continued until 1953, resulted in only 822 units. Mary bought the LH in 1958 for about 1,500 dollars. In 1962, on her way to Fort Bragg, she broke the front wheel spindle and almost ran off a bridge in the process. Emergency repairs got her back home, but that was the end of the road, as far as the truck was concerned—it has been sitting there ever since.

Next page
The photograph on the following spread shows further evidence of Mary Hitchings' obsession. The trucks in the first row are LJs from the late forties and early fifties. In the background one can catch a glimpse of an LT radiator, unfortunately with its bulldog mascot missing. These veterans of hard labor, now standing seemingly lifeless in the dense fog, have most of them been brought back from California's fertile San Joaquin Valley, a valley not only rich with regards to the crop grown there, but with respect to the variety of nostalgic machinery one can find in its wrecking yards and back lots.

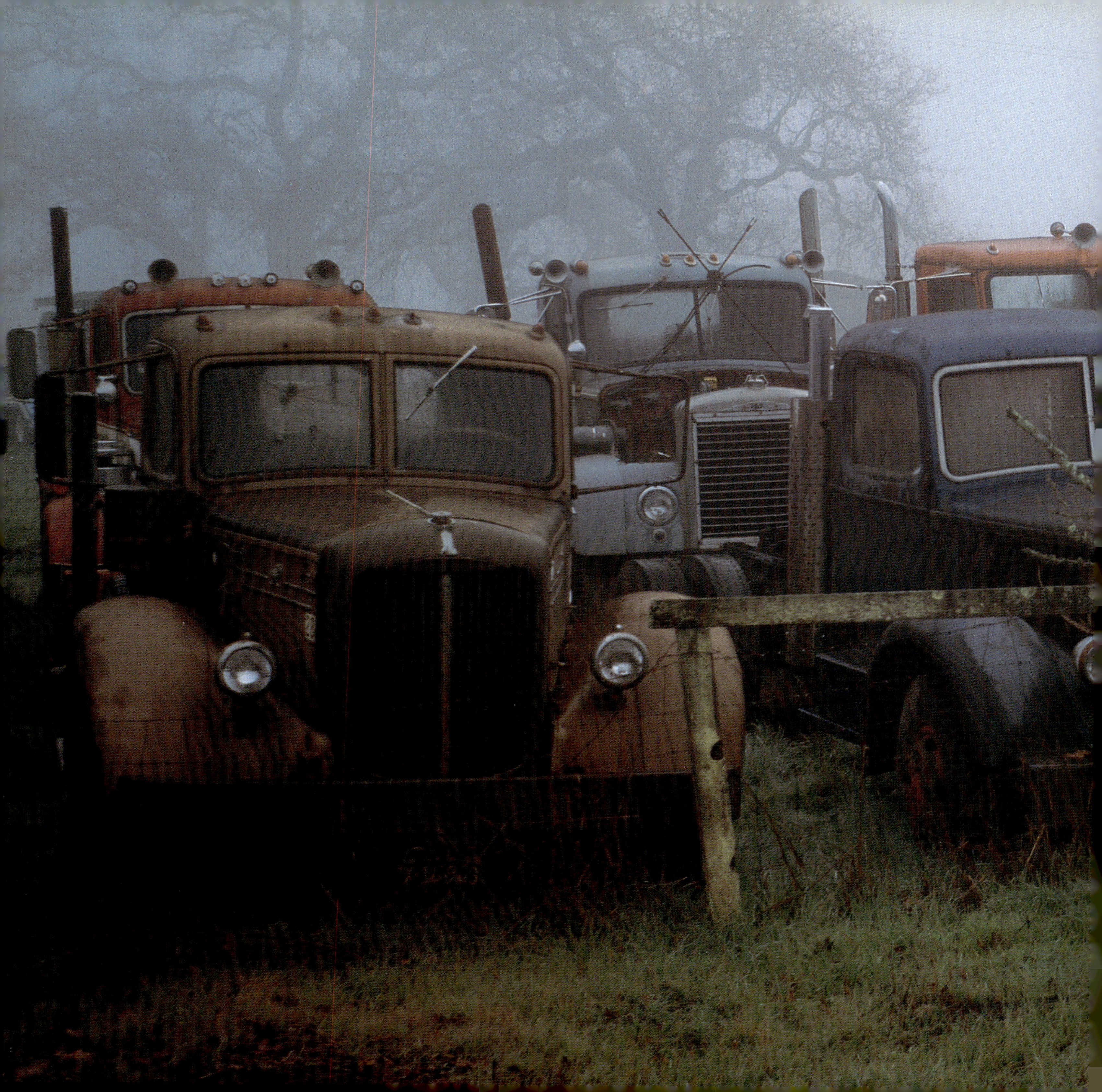

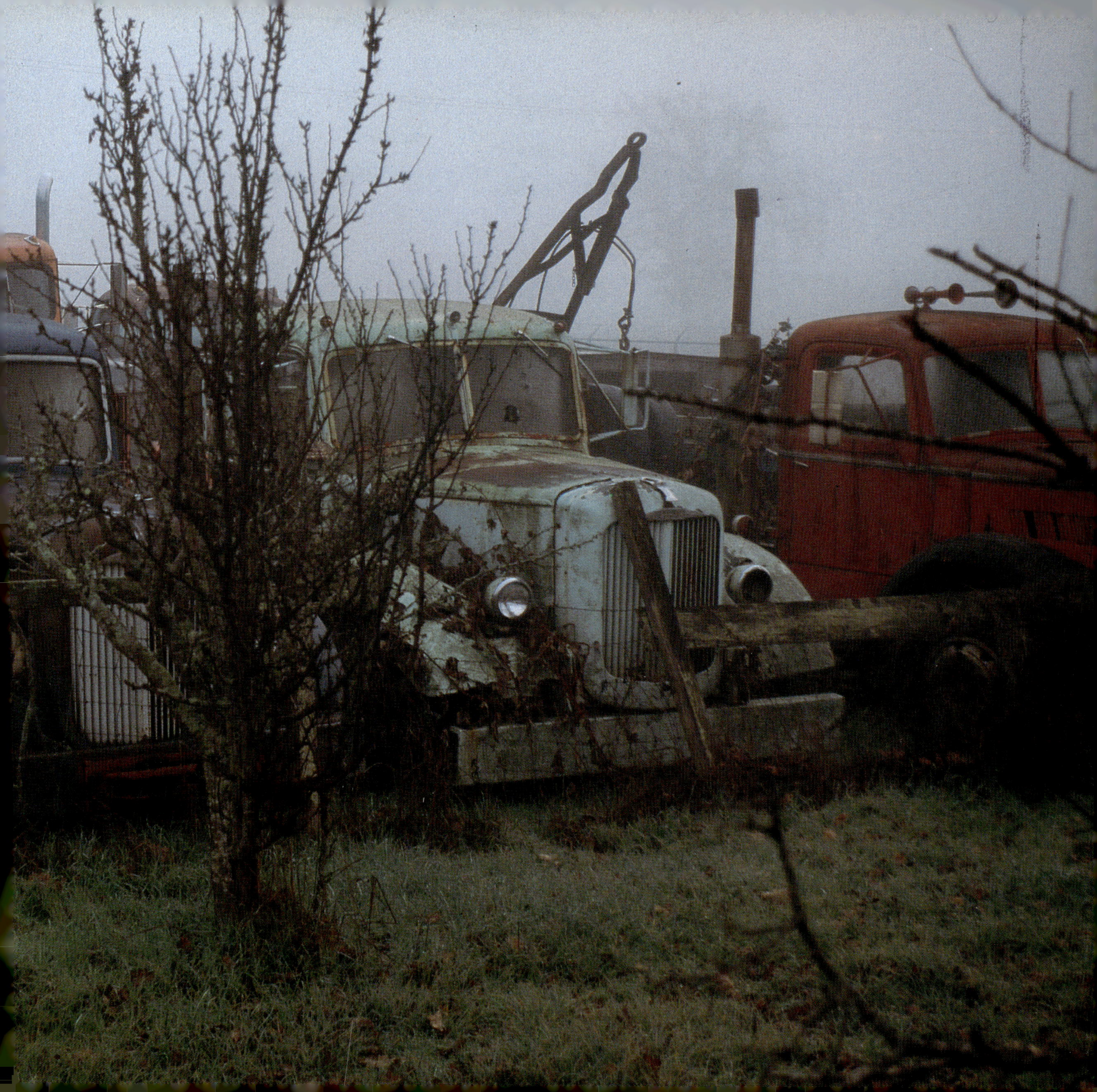

Working Mack

Hauling the loads of the land

Already from the outset the rugged product of the Mack company seemed predestined to play a leading role in the building of the nation. It was soon involved in such heavy-duty assignments as the construction of the New York subway system. Later, giant Macks helped build the Hoover Dam. Both world wars saw Mack trucks perform their part in the defense of freedom, the Bulldog and the Mack Prime Mover becoming legends in their own right. The off-road trucks and the super-heavy dumpsters have also been traditional Mack markets, as have the fields of fire engines and buses. Mack even built locomotives at one point, although this end of the business never really caught on. These sidelines of Mack activity would be impossible to cover adequately within the format of this book, which naturally must set a limit to its number of pages. Today Mack trucks of all ages, from the two-and three-decade-old workhorses to the latest products emanating from the active minds of the Mack engineers, are familiar faces along the nation's highways and byways, the following pages giving a taste of the scenery.

The R-series Mack, a sample of which is captured at speed on one of the many freeways criss-crossing the Los Angeles basin, was introduced in 1965. The versatility of this new line is illustrated by the fact that both gasoline and diesel power was offered, as was a choice of inline 6-cylinder and V-8 configurations with a variety of outputs ranging from 165 to 255 hp.

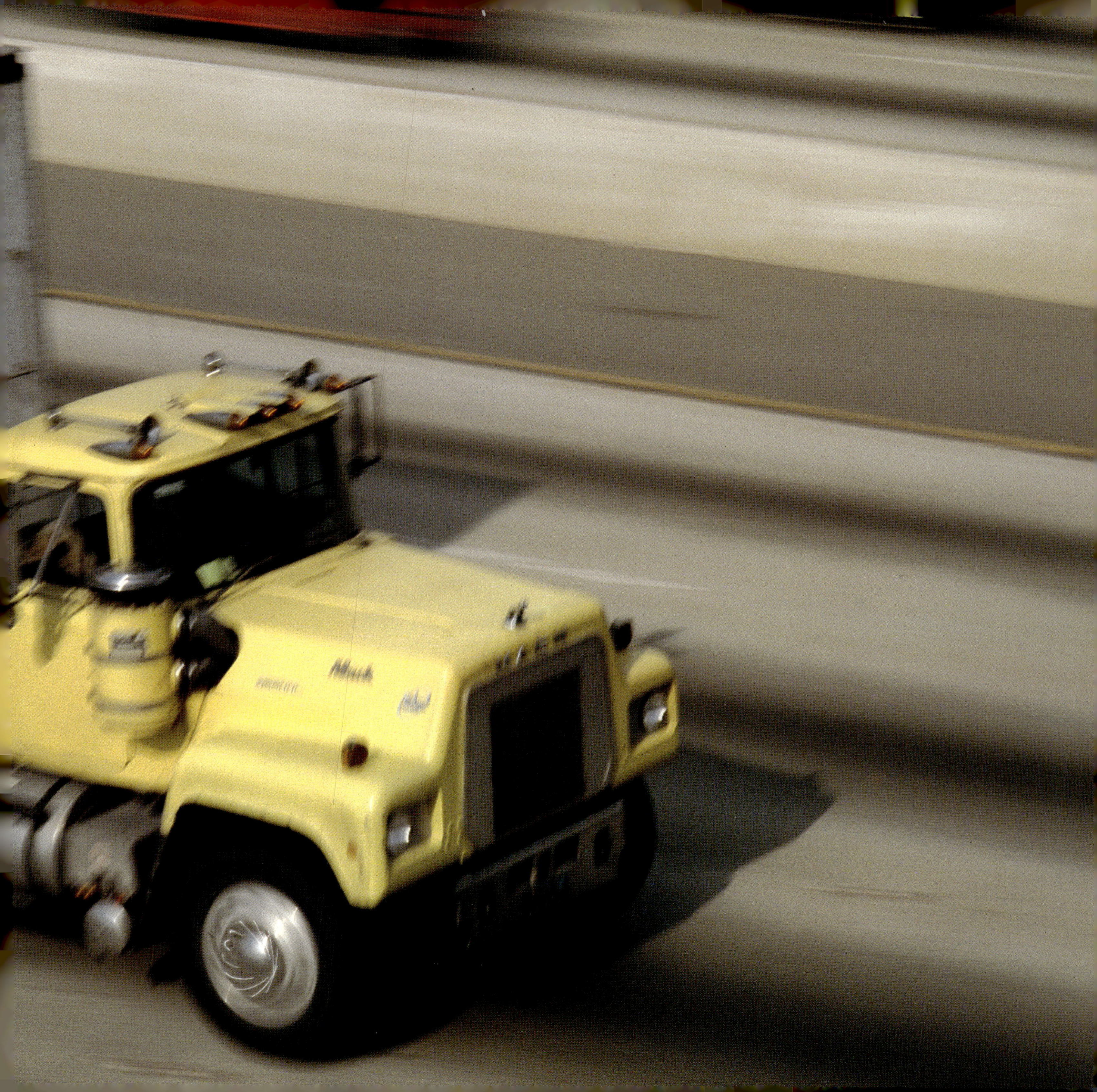

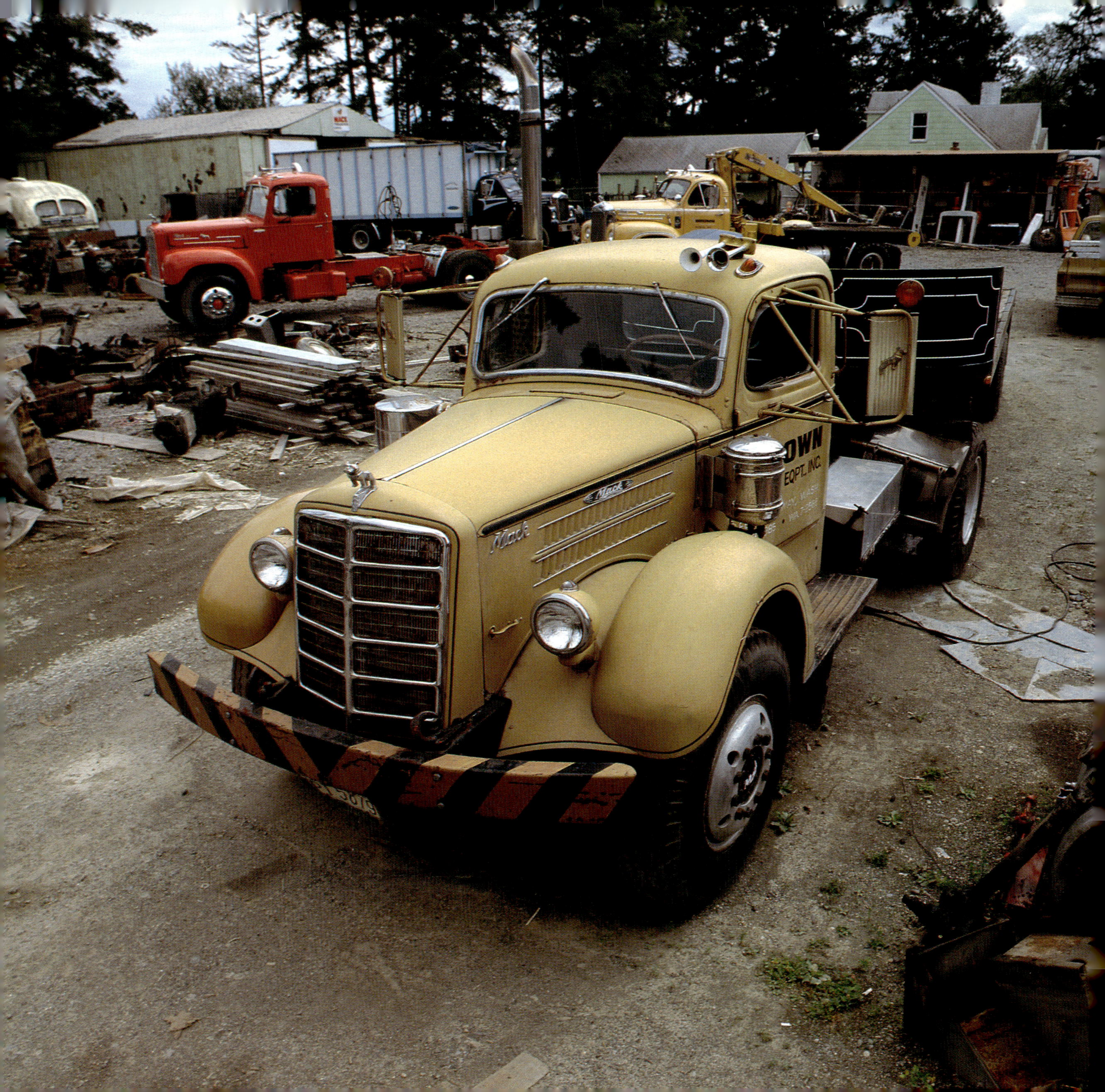
Mack
Mack
EQPT. INC.

The old Macks featured on this spread and found in Federal, Washington, are still being worked from time to time. Seen in the photograph to the left is an EQ tractor from 1948. This old workhorse is equipped with Mack's early 510 ci diesel, commonly referred to as the "Mack-Lanova" and so named because of the system of combustion chamber design chosen by the Mack engineers. In the background can be seen three additional Macks, all in the classic B-series.

Pictured in the photograph above is an H-series cabover tractor from 1961, formerly used by a van line. Introduced in 1952, this series of Mack trucks was characterized by the short distance from bumper to back, which made it possible to stay within the 45 foot overall length limit and still carry a maximum of load. Mack's own Thermodyne diesel or gasoline engines were fitted, giving the type a 170 hp rating. The H-series came in 18 sub-configurations, some of these were still in production as late as 1966. A total of 12,296 units were built.

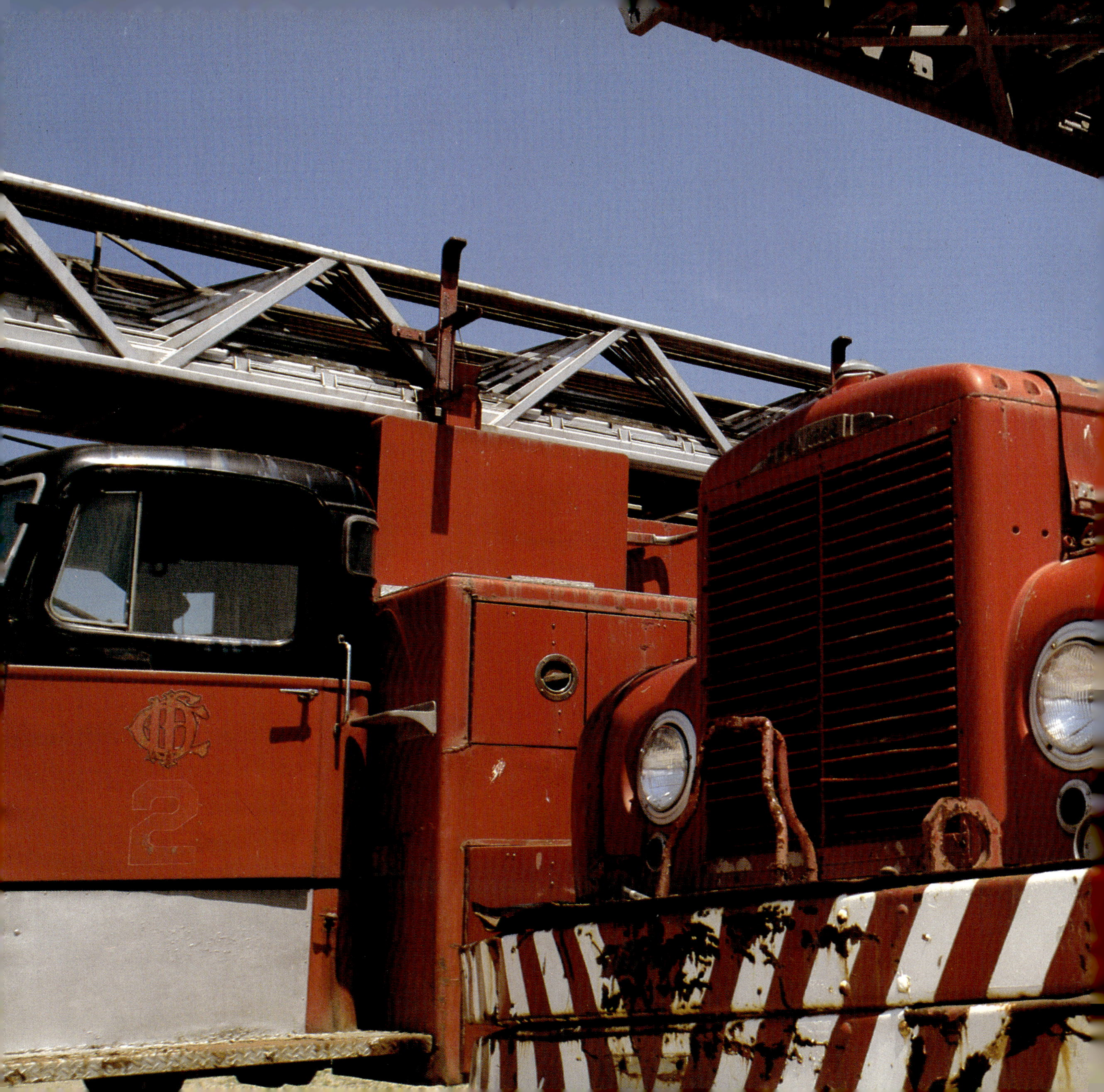
2

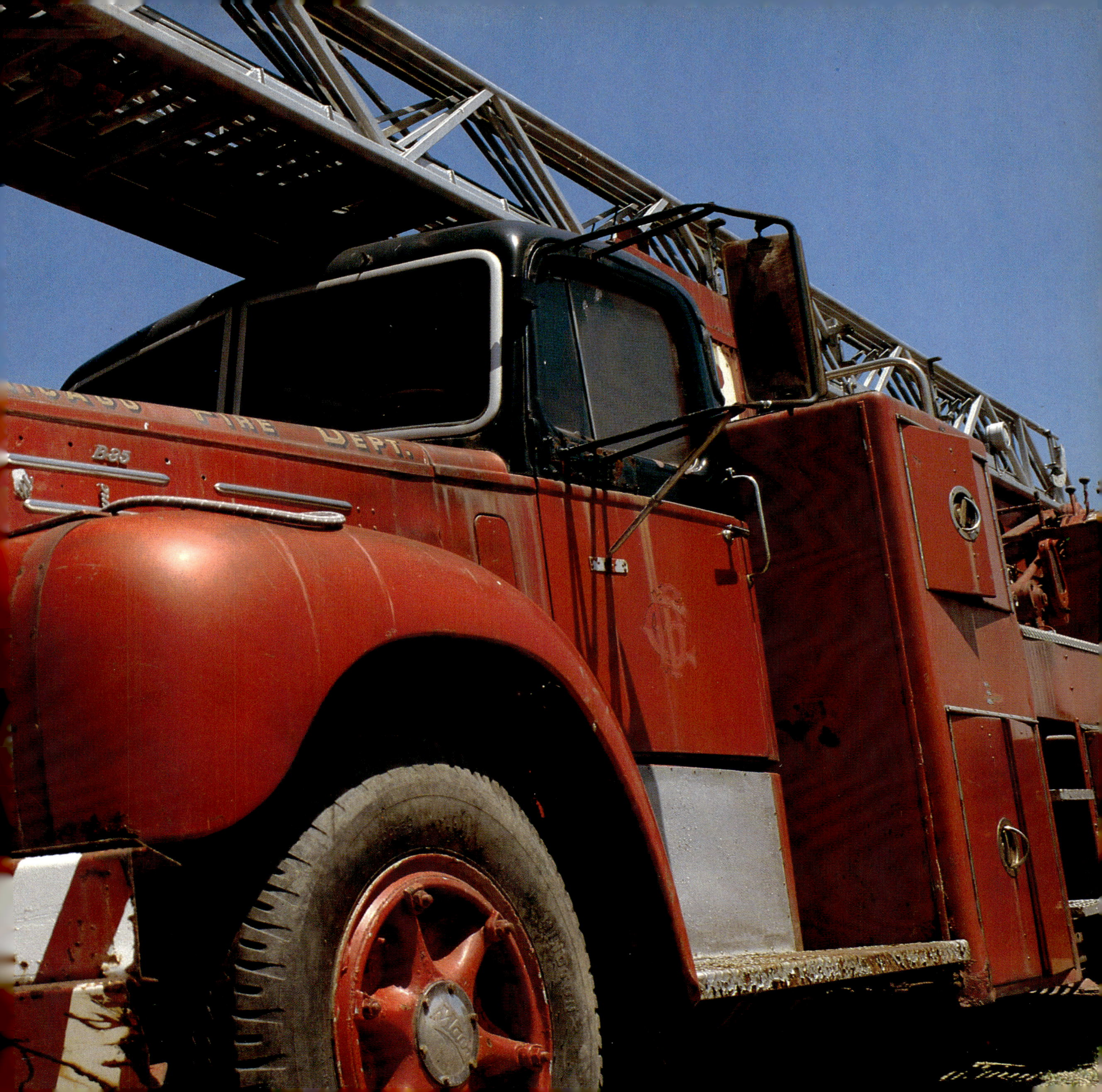
FIRE DEPT.

Previous page
The photograph on the previous spread shows the biggest of the firefighting breed, the B-85, this particular example being one of about half a dozen units purchased from the Chicago Fire Department by Manders Diesel Repair in Lakeville, Minnesota. A total of 77 B-85s were built between 1956 and 1964. Pictured on this spread is a close-up view of the engine compartment of the B-85. Power comes from a 707 ci gasoline unit of Mack's own manufacture, producing around 200 hp. The pumps varied in size from 500 to 1,250 gpm.

Mack trucks soon became popular for use as fire engines, and the company ultimately built up a special division to serve this sector of the market. By 1953 more than 1,000 local fire departments across the nation were committed to Mack equipment. In 1957 it became necessary to enlarge the operation and Mack moved the division to a separate 200,000 square foot plant in Sidney, Ohio. Many fire departments still run their old Macks. Shown here is a Pirsch fire truck with a Waukesha motor.

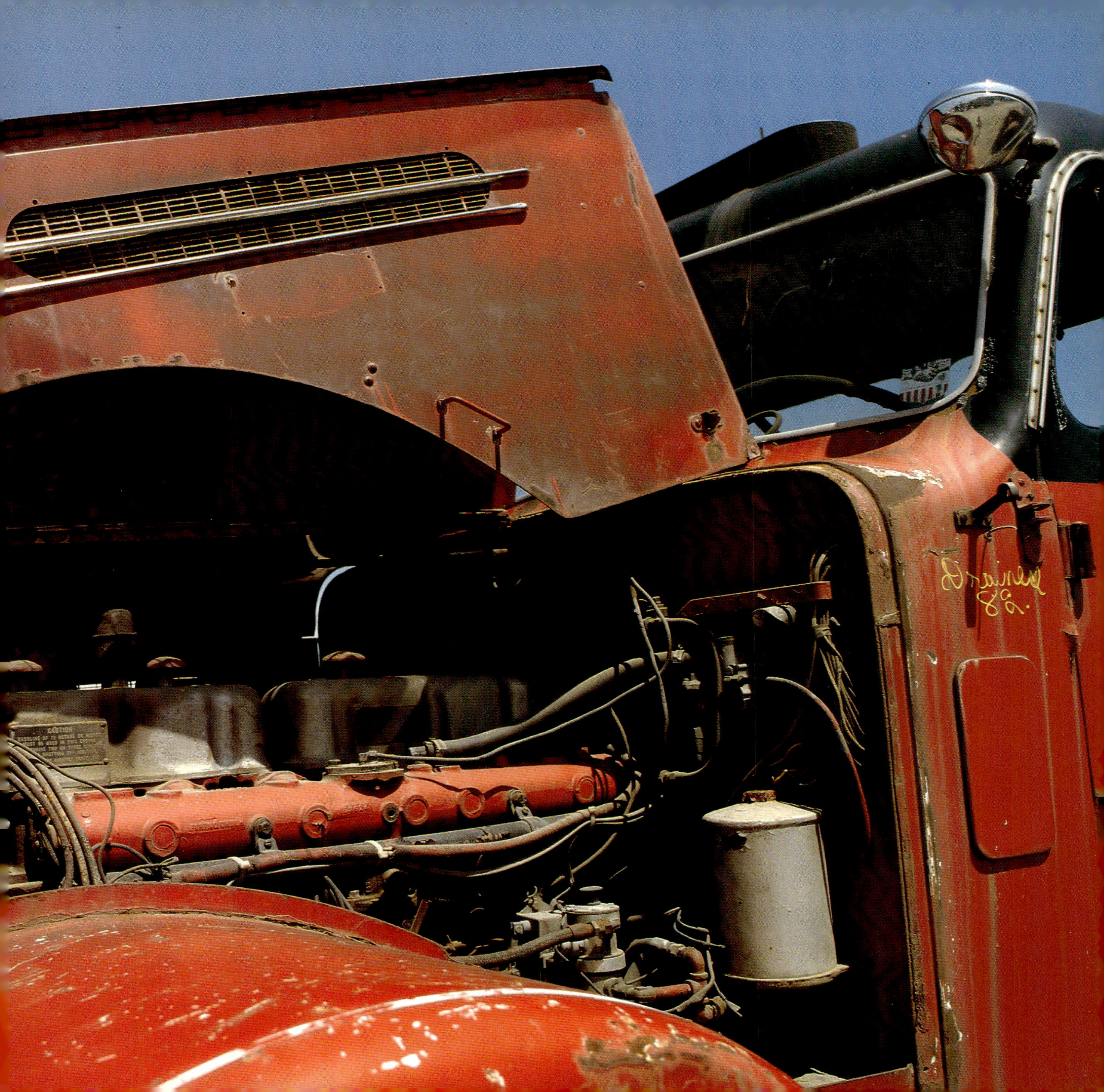

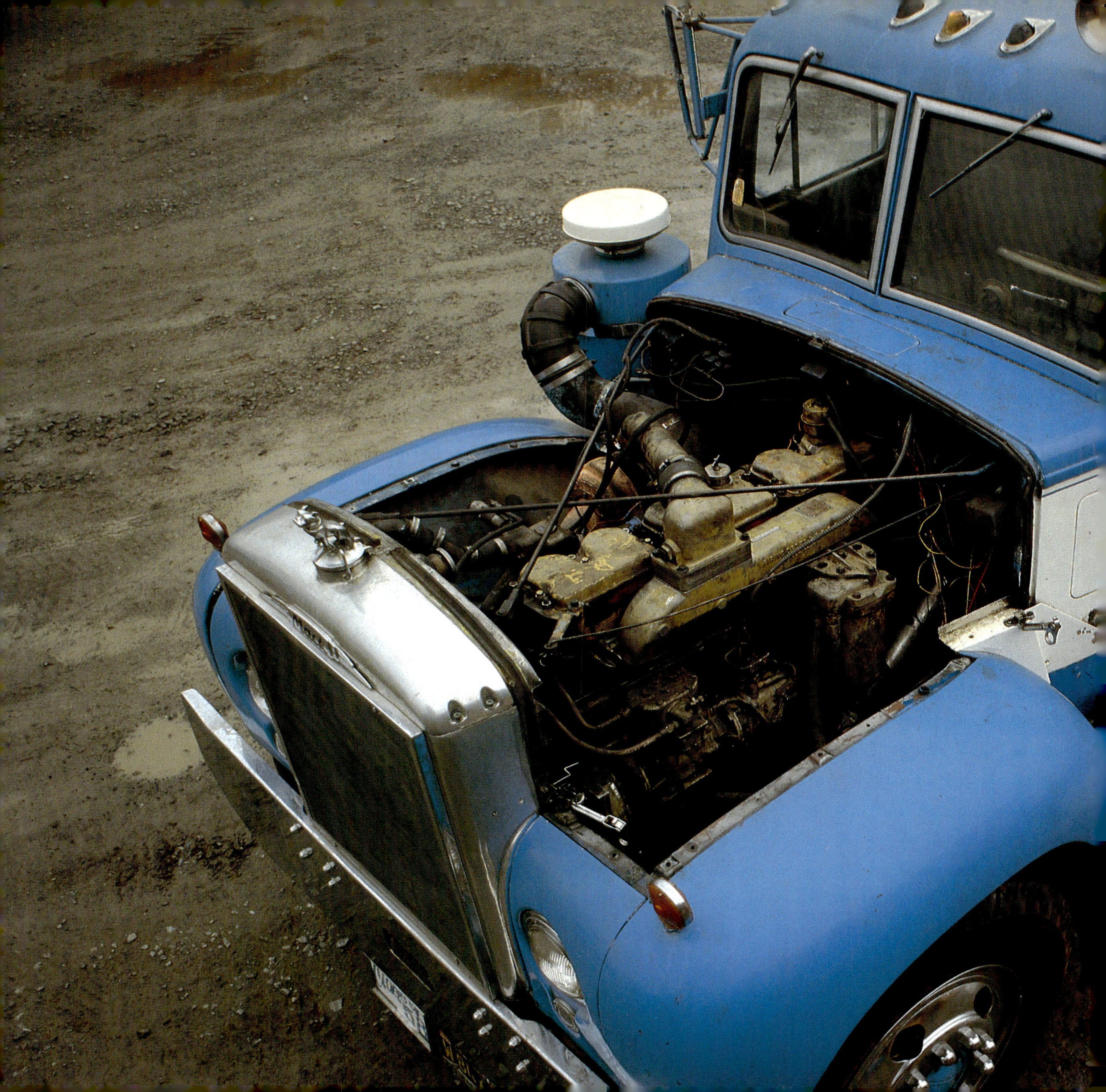

The heart of this B-73 from 1958—still used to haul lumber—is a 743 ci Cummins diesel, producing 262 hp. The large cylinder located on the passenger side of the cab is the air cleaner. From there the air passes through the turbocharger, placed on the same side, crosses over the engine and enters the intake manifold, seen on the right-hand side. Fuel is carried in two exterior tanks, located on either side of the cab, below the doors. They hold 55 gallons each. The small cylinder on the driver's side, just ahead of the fuel tank, holds the oil filter. A total of 2,520 B-73s were built between 1955 and 1956.

The services of the converted old wrecker seen in the photograph on this page, are still appreciated by the operators of a truck repair outfit in Minnesota, who finds its awesome pulling capability just the right ticket for their heavy-duty assignments. The B-61, which is of a 1959 vintage, is powered by Mack's 673 ci Thermodyne diesel. A total of 15 gears helps disperse the power in the proportions required by this kind of work, which seems to have become a specialty of old semi-retired Macks.

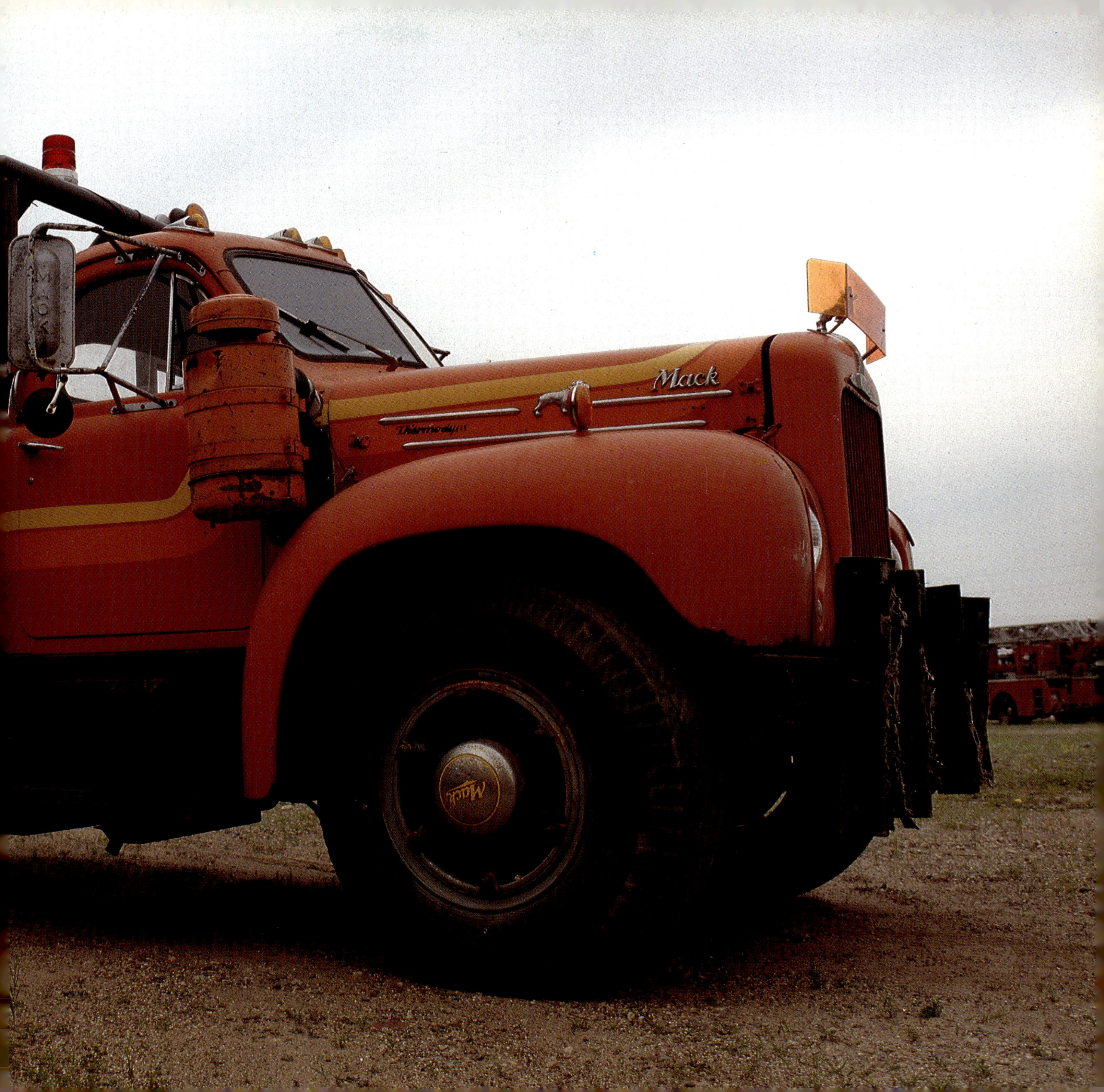
Mack

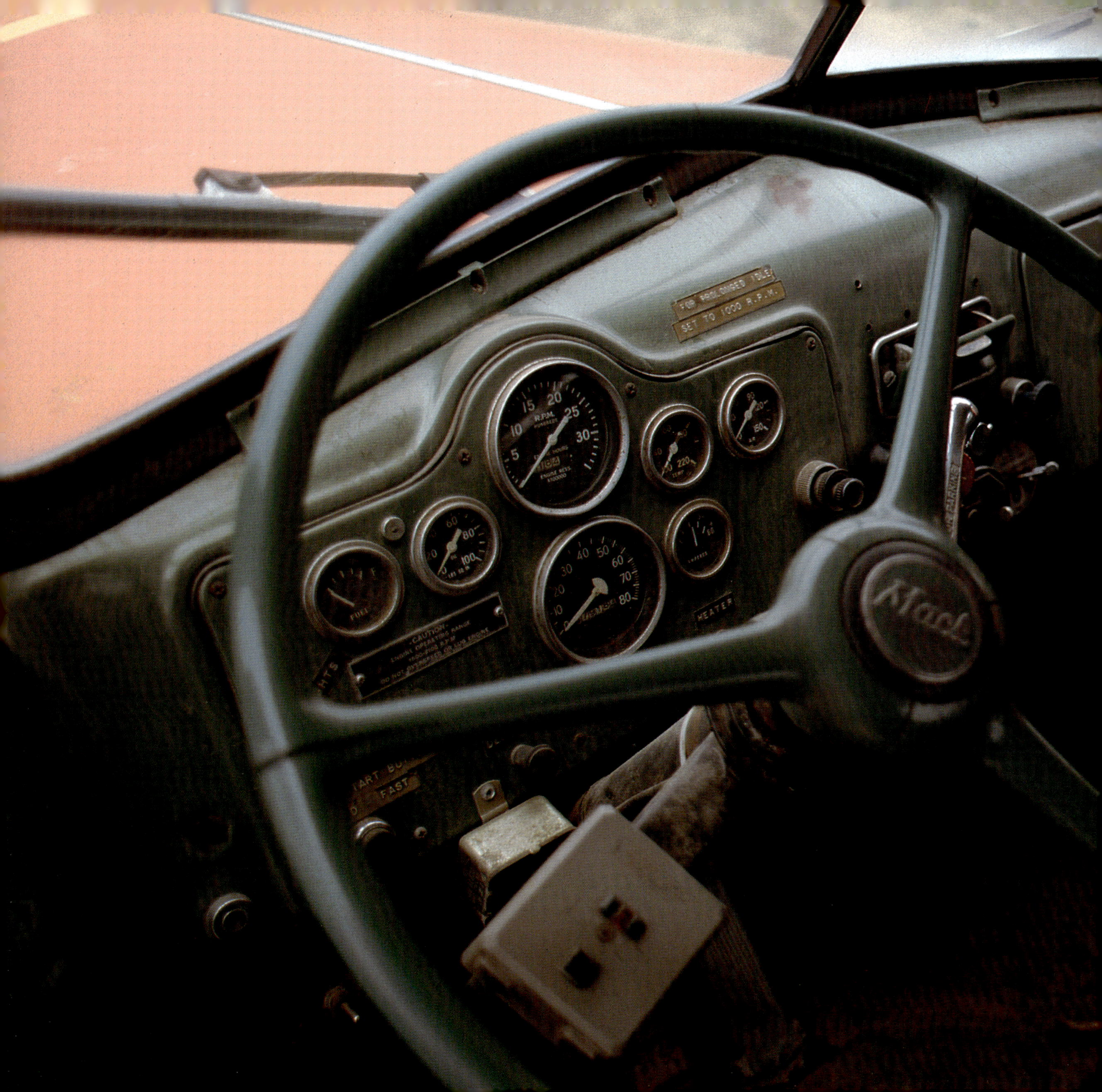
FOR PROLONGED IDLE
SET TO 1000 R.P.M.
R.P.M.
FUEL
HEATER
FAST
Mack

Seen in this photograph is the cockpit of the B-61 wrecker featured on the previous spread. The instrument console includes an impressive array of gauges, the tachometer being the upper of the two large ones, and the speedometer being the lower. Fuel and oil pressure gauges are grouped to the left, while the temperature, air and ampere meters are found on the right. The two floor-mounted levers operate the main and auxiliary transmissions. With the addition of the large steering wheel, all these elements form a most nostalgic picture, remembered by a whole generation of truckers and truck enthusiasts.

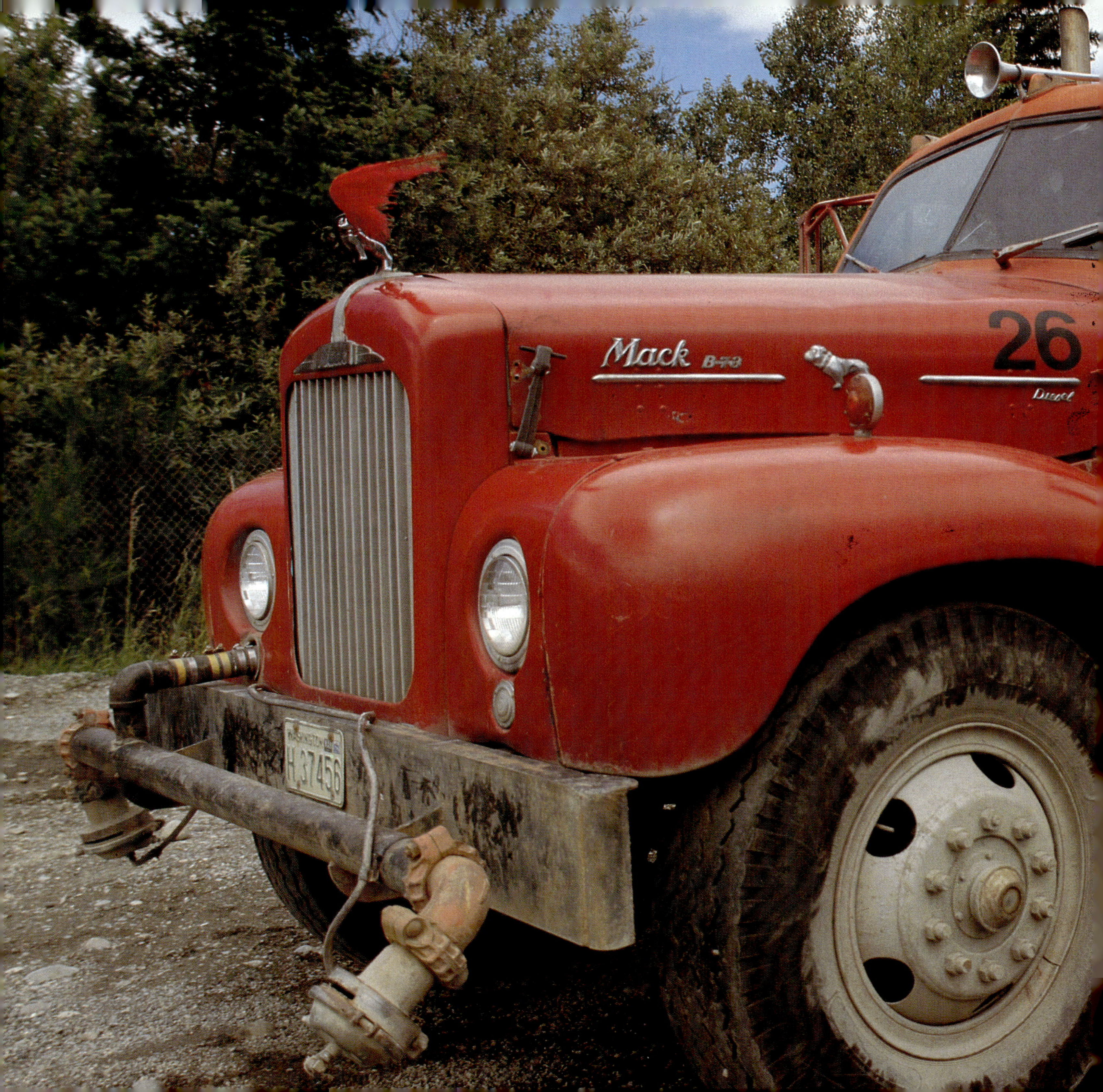
Mack
26
H 37456

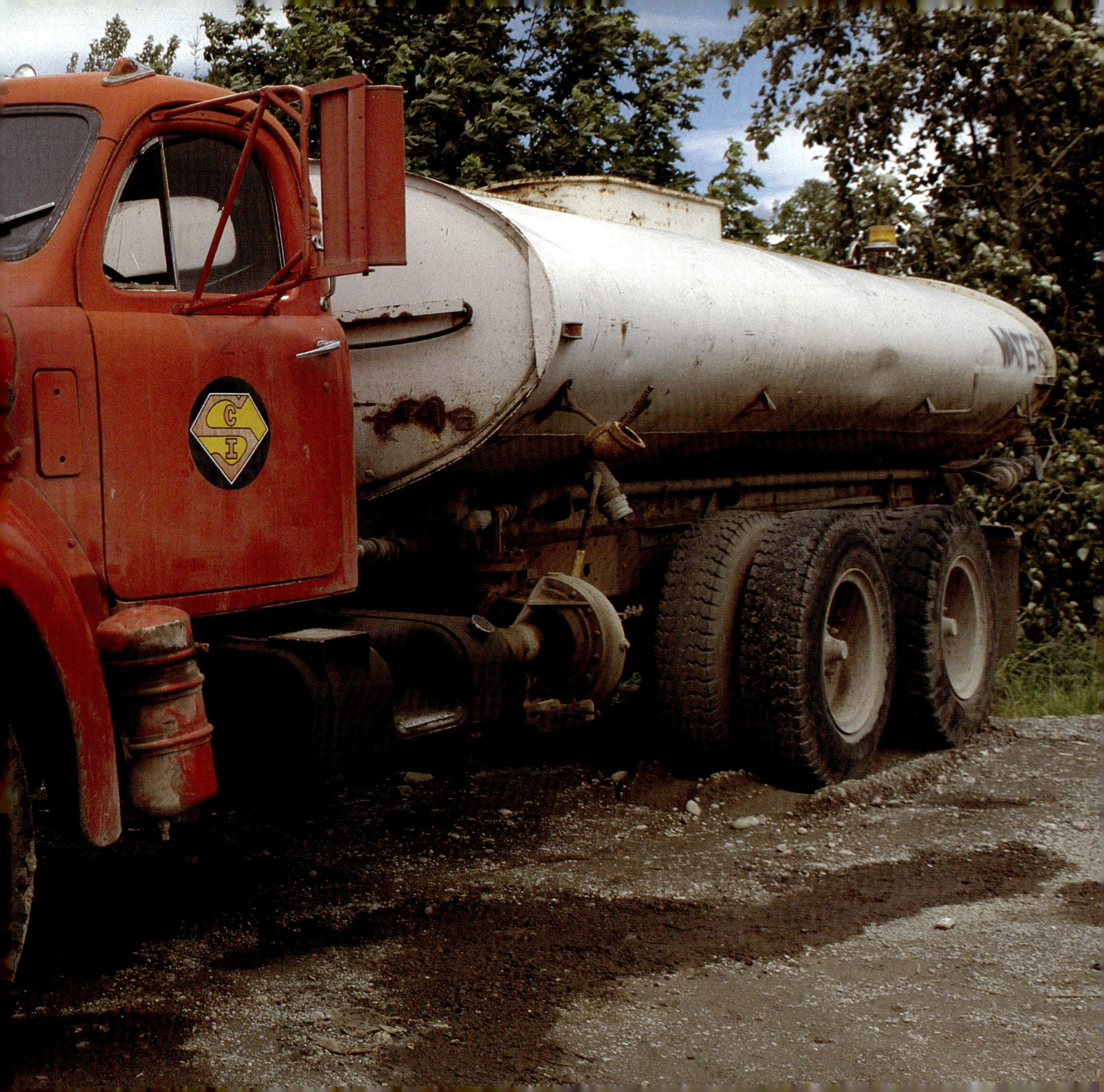

Previous page
Pictured on the previous spread is another favorite occupation of old Macks. This B-73 was photographed in the state of Washington, where it was used as a water truck on a large road improvement project. Its owner is apparently a Mack enthusiast equipped with a certain amount of humor—notice the proud red ostrich feather decorating the bulldog mascot. There seems to be no end to the longevity of a Mack, and, regardless of its condition, a Mack still possesses an aura of magnificence.

The photograph on this page is yet another illustration of the longevity of Mack machinery. The owner of this B-series Mack from 1964, operating out of Scandia, a small town located right on the border between Minnesota and Wisconsin, does not give up his Mack wheels easily. And the reason should be obvious. A rig like this—no matter how battered it looks—with its multitude of wheels and its magnitude of bed footage, is a valuable asset in the farm country, where feed and fertilizer must be moved with the greatest of efficiency.

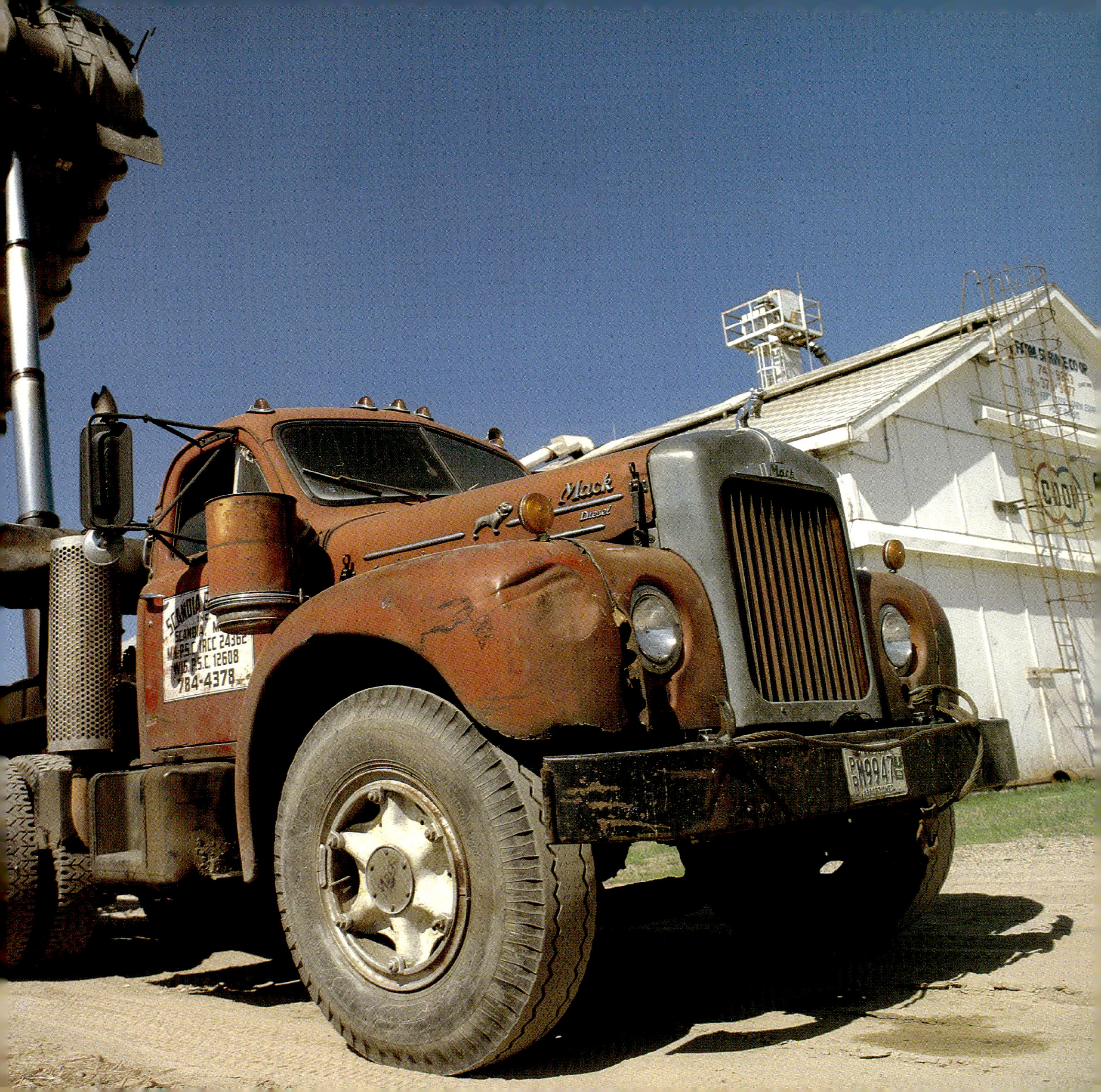
Mack
Diesel
WIS P.S.C. 12608
784-4378

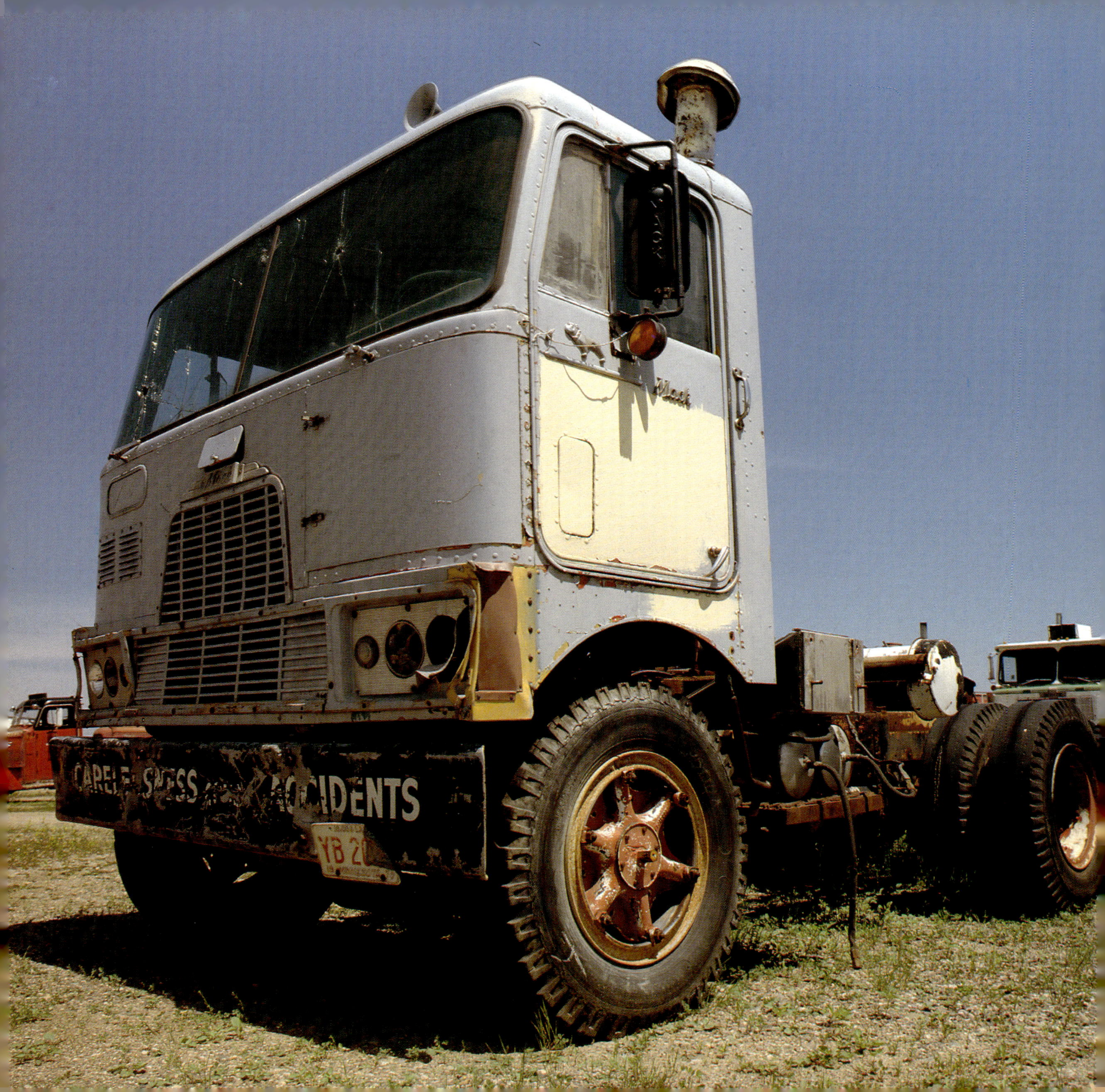
Mack

Featured on this spread are two extremely rare workhorses. To the left, a G-75 from 1960. Introduced in 1959, the G-series included seven sub-models. A total of 2,181 units were produced, whereof only 319 were G-75s. Production ended in 1962. This cabover tractor featured a narrow cab, following the industry norm referred to as BBC, which indicates the distance from the bumper to the back of the cab. Power came from either Mack or Cummins, with the latter boasting a maximum output of 335 hp.

Seen in the photograph above is a C-607, which had an extremely short production run of only 301 units. It was built between 1963 and 1965. Conceived as a combination of the conventional type of truck and the cab-over-engine design, the C had a distinctively snub-nosed appearance. Under that short hood hid a vendor-supplied engine or Mack's new powerful V-8 diesel, which produced as much as 255 hp without resorting to turbocharging. While the G-series had a standard cab width of 51 inches, the C-series measured 89 inches.

Guasti Rd

It seems every call for heavy-duty equipment is answered by a Mack, as seen in the photograph on this spread. The load of a trailer shifted suddenly and it overturned right in one of the busiest intersections of the trucking world, that of Terminal and Guasti in Ontario, California. The RL-700 is a 1973 model and was before the conversion a low-bed tractor. Power comes from Mack's Maxidyne diesel, developing 325 hp. The hydraulically operated Kemp recovery unit lifts as much as 35 tons.

Next page
Another type of heavy-duty vehicle where Macks seem to have cornered the market is that of the cement mixer. These Macks are of an off-road type and have beefed-up springs and roll-on flotation tires. Notice that the air cleaner is of the dry type. The particular unit featured in the picture on the following spread is one of a fleet of 1975 and 1976 models. A full load means that 27 tons of cement is carried in the mixer.

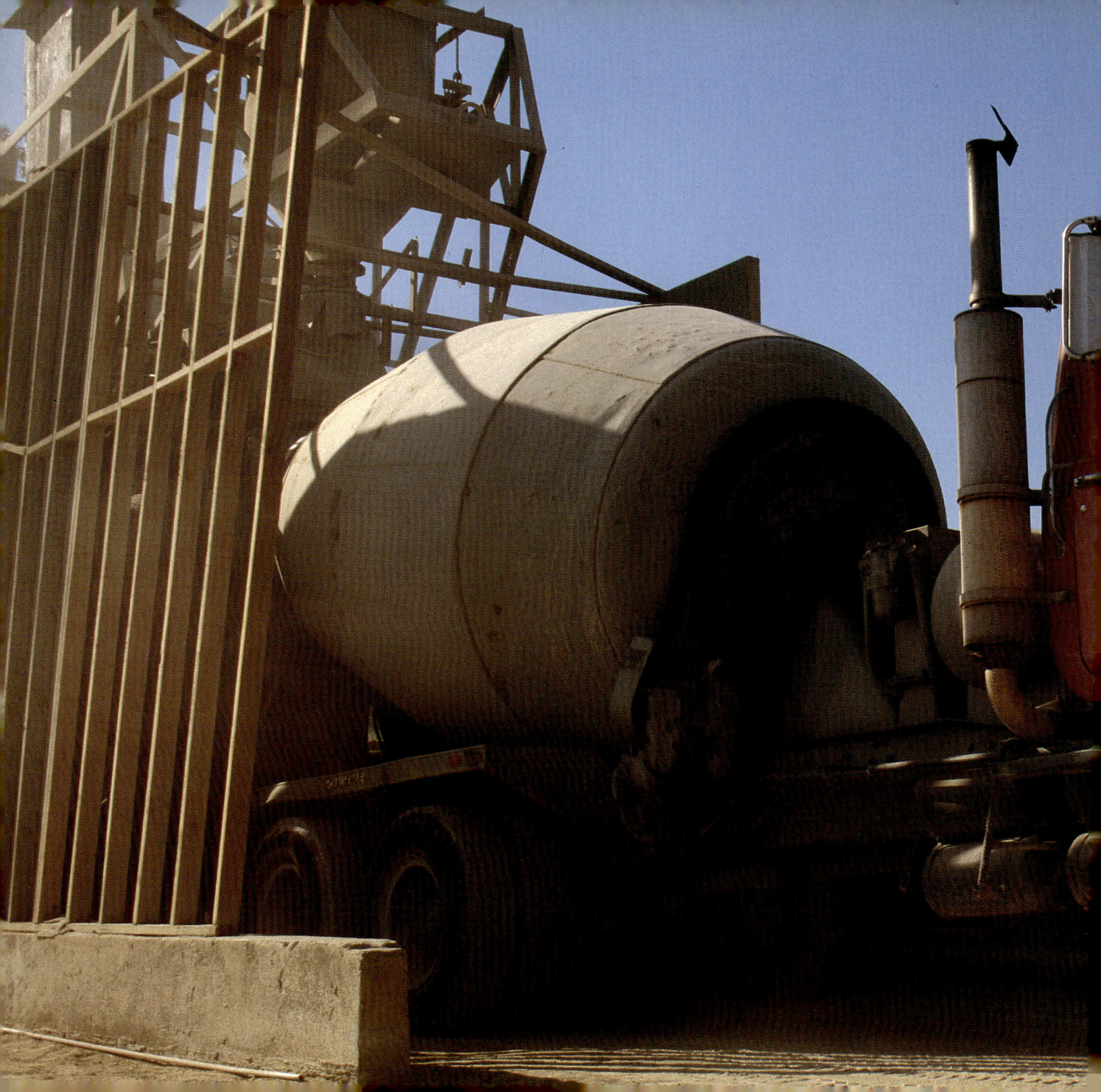

BAKER
GUNITE
&
READY MIX CO.
(714) 829-3453
FONTANA, CA

Mack

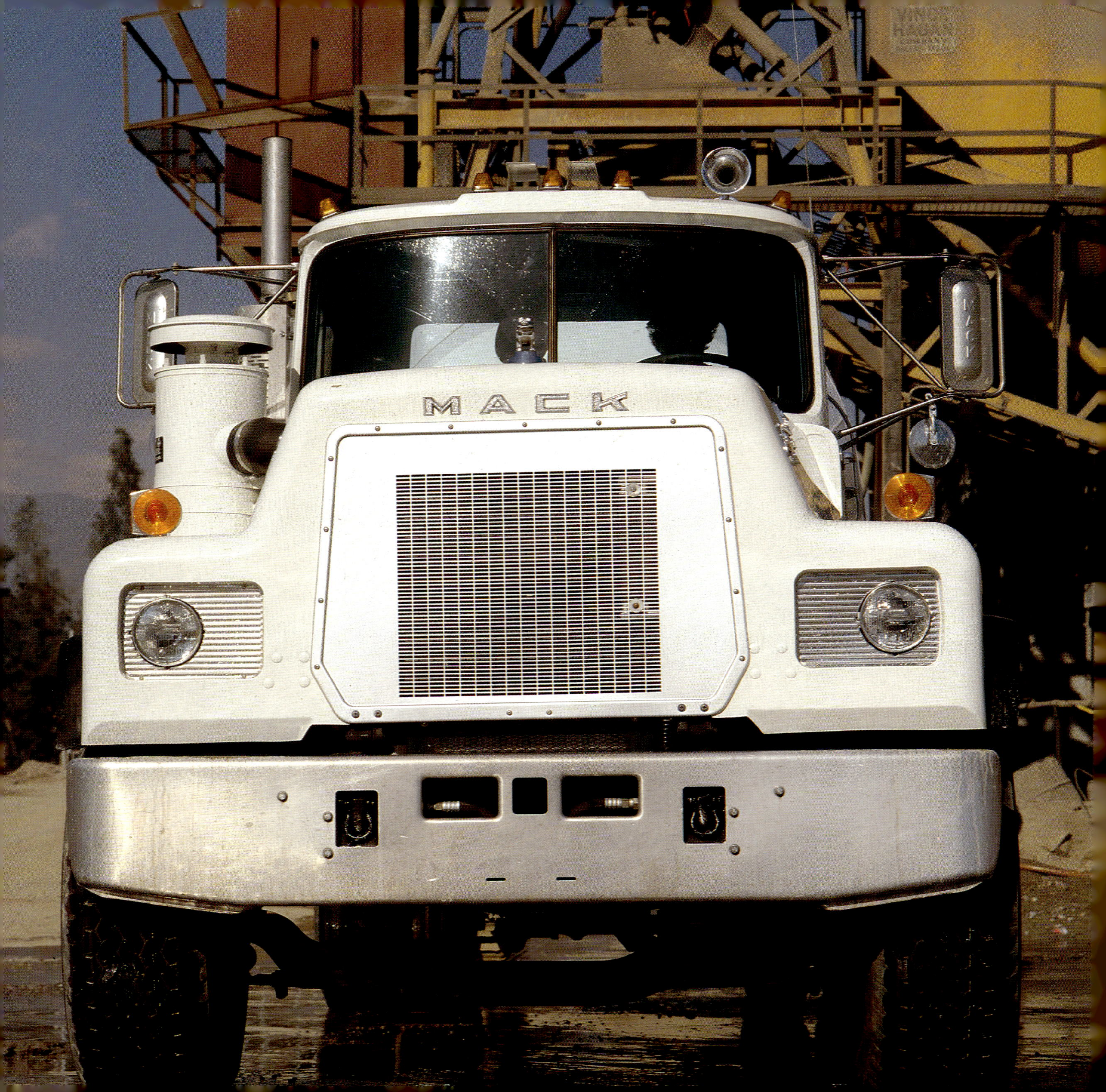
MACK

This spread captures two very familiar Mack faces. To the left is a member of the R-series. Introduced in 1965, this line of trucks captured the mantel left vacant by the B-series. With its massive, snub-nosed front, this design builds on the classic Mack look, pioneered by the Bulldog, and re-enforced by the B-trucks. However, in the R-series the design has reached an ultimate level of functional straightforwardness. It's no longer a question of beautiful or ugly—awesome is the word.

The photograph above captures the mean visage of Mack's Cruise-Liner, in traditional Mack nomenclature also referred to as the W-series. This new cabover truck was introduced in 1974 and brought truck technology forward in a giant stride, with important innovations in frame and suspension construction as well as in the area of creature comfort. The Cruise-Liner featured a wide variety of engines, both proprietary, such as the popular Maxidyne, as well as of vendor origin, such as Cummins, Detroit Diesel and Caterpillar, with outputs reaching as high as 440 hp.

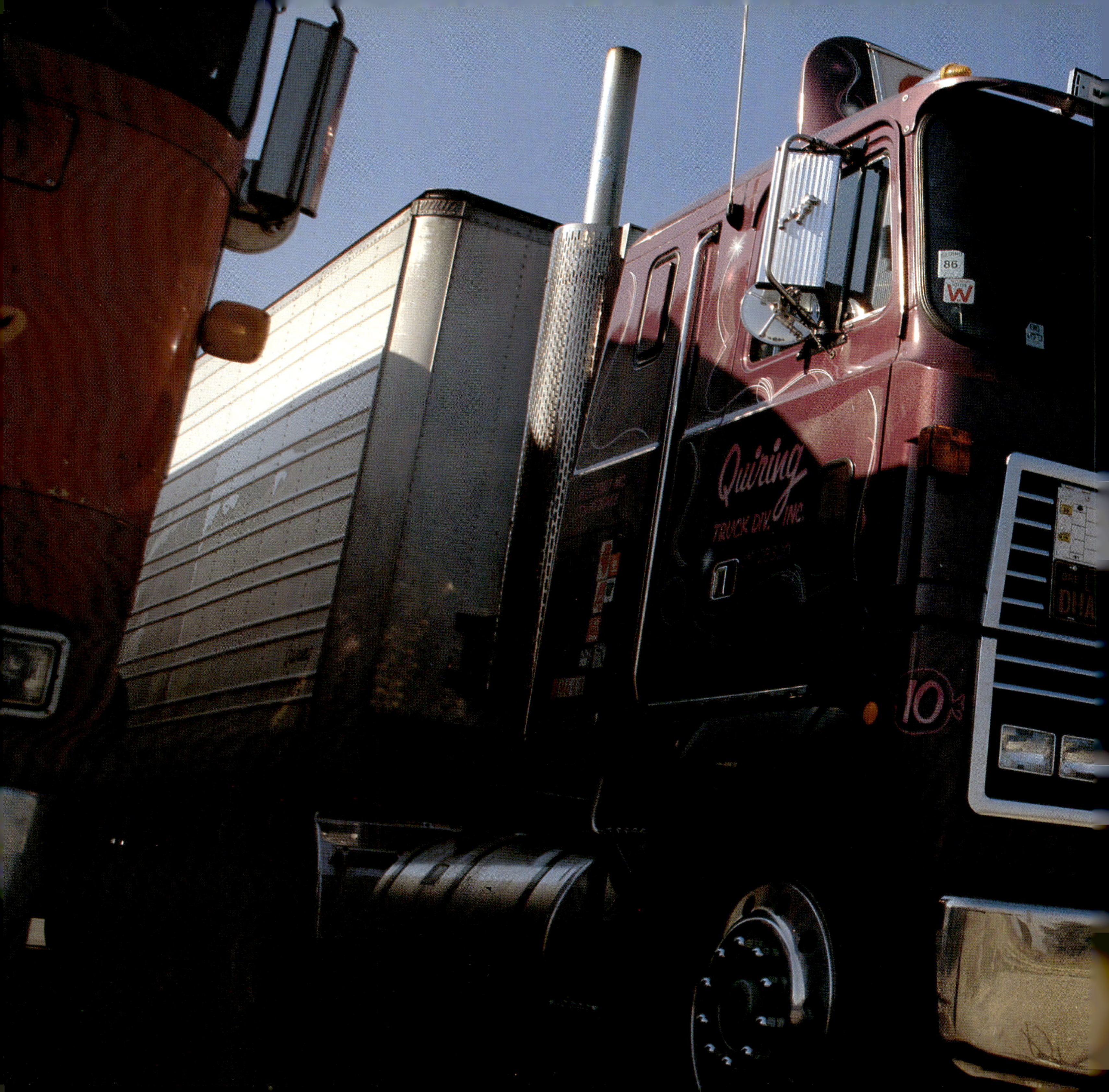
Quiring
TRUCK DIV. INC.

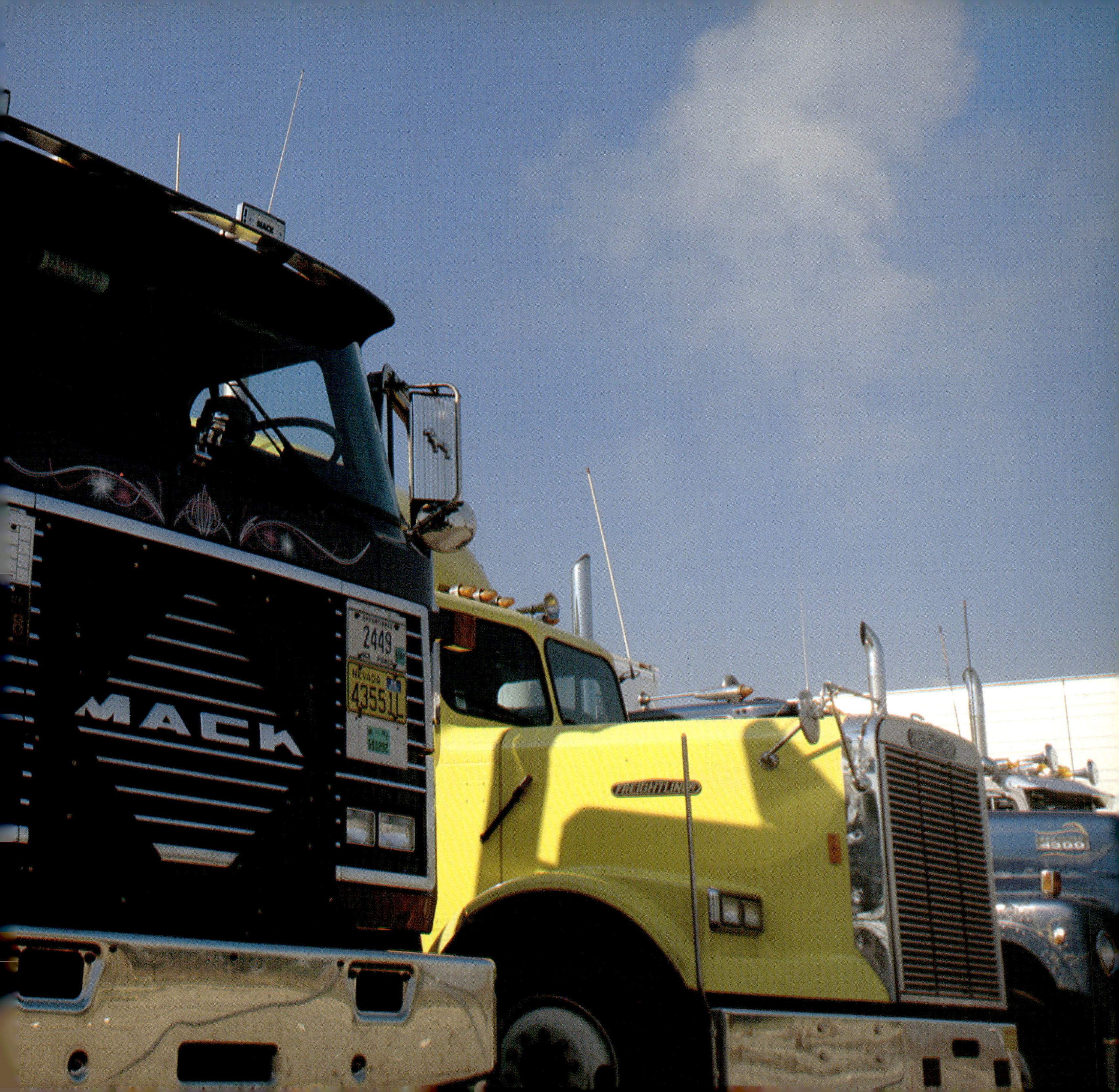
MACK
FREIGHTLINER

Previous page
The photograph on the previous spread shows a member of the newest-generation Mack, a 1986 Ultra-Liner. This cabover line is the result of many years of research, where both aerodynamic experiments, safety considerations and technical innovations were brought together to produce one of today's most outstanding trucks. Here the massive giant is parked alongside its peers in a Southern California truck stop, while the driver takes a nap in the sleeper unit. Even in this latest creation, the old bulldog is proudly carried up front—never at rest, always ready to go.

The photograph on this spread shows another member of a relatively new type of Mack trucks, the Mid-Liner, here seen in the service of a Southern California meat packing company. This new type of truck, introduced in 1979, was created to fill the need in the field of medium-sized delivery-type vehicles. Built in France, by Renault, it has Mack-specified features such as new technology air-to-air aftercooling and turbocharging. Diesel powered, the Mid-Liner comes with a variety of engine sizes, varying in output from 125 to 175 hp.

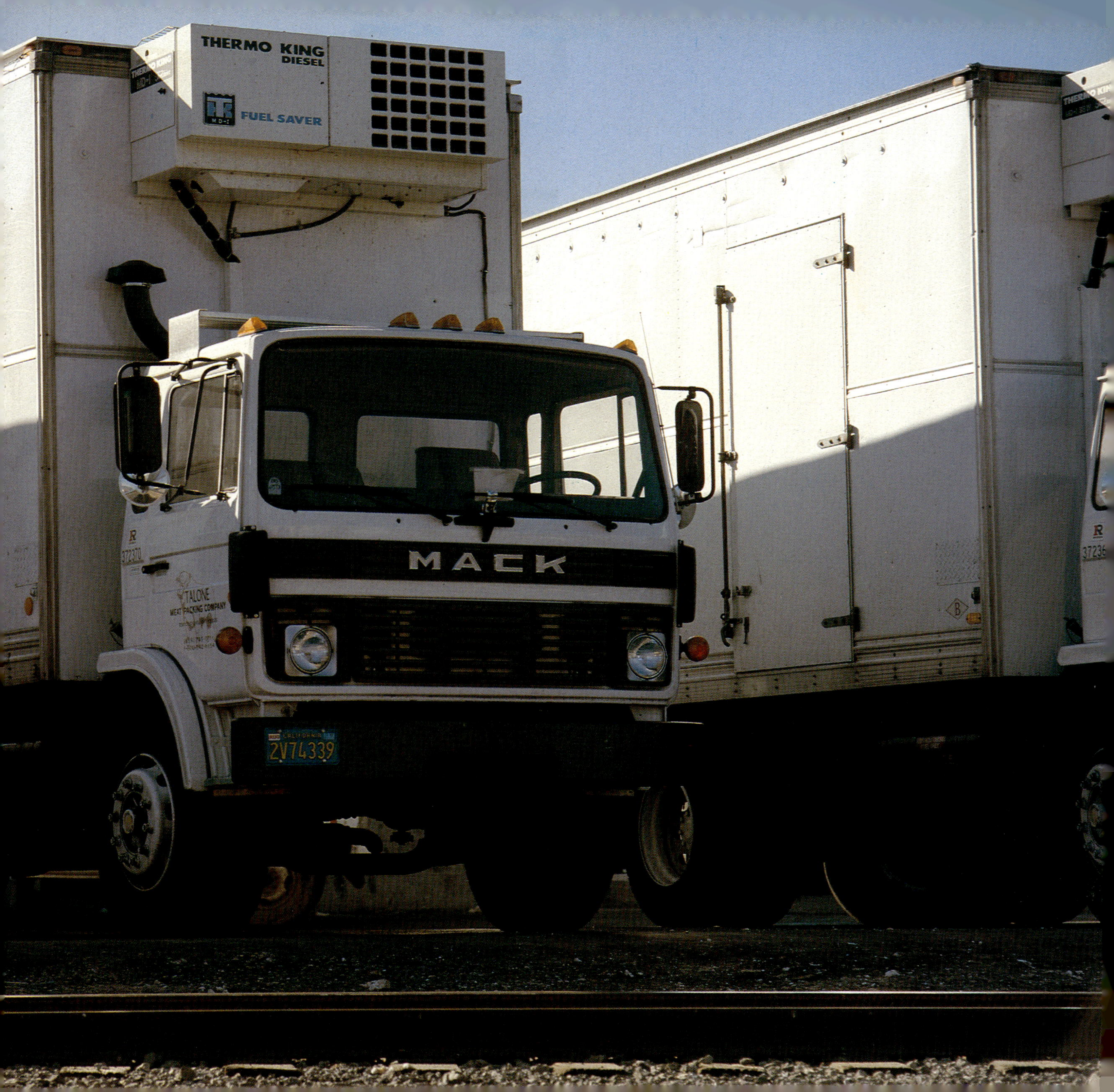
THERMO KING
DIESEL
FUEL SAVER
MACK
2V74339

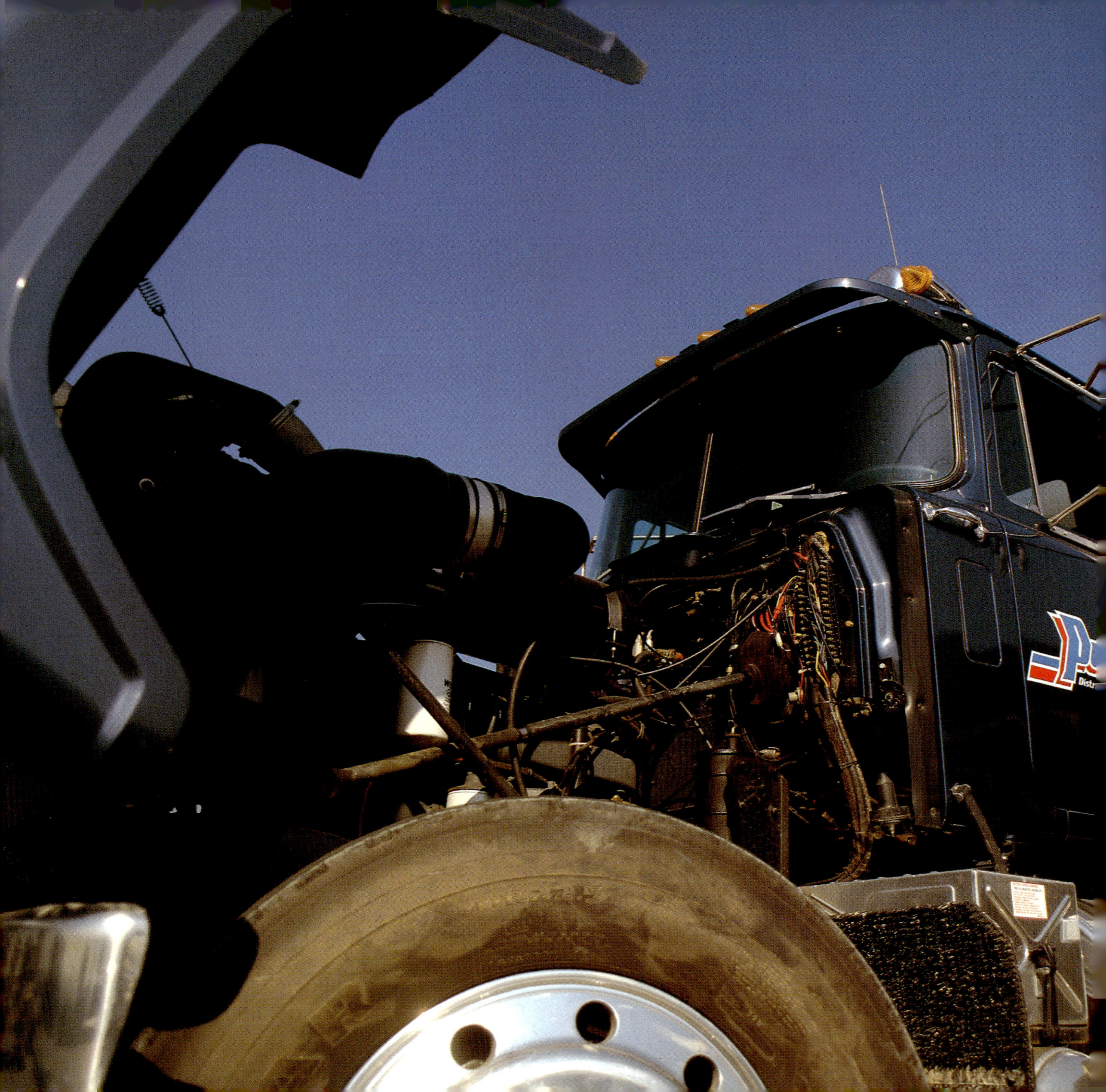

One of the most awe-inspiring faces seen on today's highways belongs to Mack's Super-Liner, introduced in 1977. While the cab is of the design proven in the old R-series, manufactured from galvanized steel, the hood, with its conventional-style radiator enlarged to enormous proportions, is made from fiberglas and tips forward for easy access. Power comes from Mack's own turbocharged and air-cooled in-line-six diesel, the Maxidyne, developing 300 hp, or the Econodyne V-8, unleashing as much as 500 hp.

Next page
The photograph on the following spread features the frontal aspect of Mack's Super-Liner. The size of the grille, an impressive 1,500 square inches, is necessary in order to facilitate the large quantities of cooling air required by today's enormous engines. When first introduced, the Super-Liner featured round headlights, placed side by side in groups of two. On the Super-Liner II, introduced in 1985, square halogen-type headlights were fitted, which further improved the purposeful expression on the face of Mack's flagship for the eighties.

TESTED H V 12 85

Poma
Distributing Company, Inc.

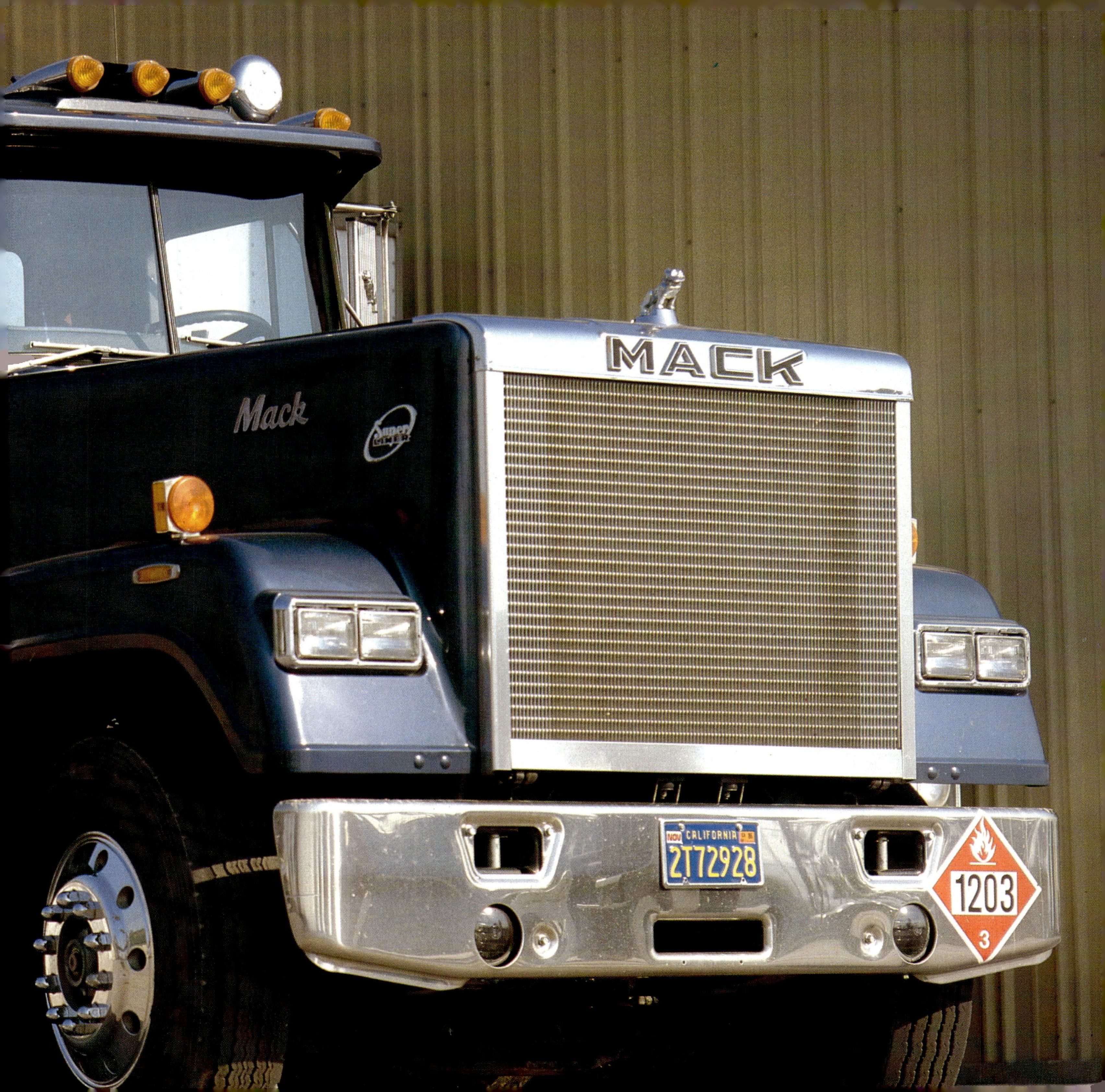
MACK
Mack
NOV
CALIFORNIA
2T72928
1203
3

Modern Mack

High-tech raiders of the road

The old phrase "Built like a Mack" is as appropriate today as it ever was. Although it has always been the goal of Mack engineers to build not only strong, but light, this philosophy has never been as refined as in the modern Macks. For instance, the revolutionary new drop-section frame is lighter than any other straight-rail frame with the similar ride characteristics. In addition, low rate, no-maintenance taperleaf springs offer vastly improved suspension and increased carrying capability. Inside the cab, air springs and shock absorbers have increased comfort, as has the top-of-the-line option of posh red velour interior. Add to this the luxury of the new sleeper boxes, available in widths from 36 to the walk-in 60 inch size, which comes complete with hanging closet and its own heating and air-conditioning system. Engine technology has certainly not lagged behind; charge air-cooling, turbocharging, and four-valve heads all spell an increase in power and fuel efficiency—and again, weight saving, to the tune of 600 pounds over competitive power plants. Built like a Mack—still.

The ultimate Mack is the subject of the photograph on this spread. Recently, engineers, stylists and marketing men put their heads together and came up with the new limited-edition Mack Magnum, a Super-Line specially geared toward the owner/operator. The already-powerful Mack V-8 Econodyne has received another 50 horses, to make it one of the most powerful on the road. Also pretty powerful is the price tag of 75,000 to 100,000 dollars, depending on equipment.

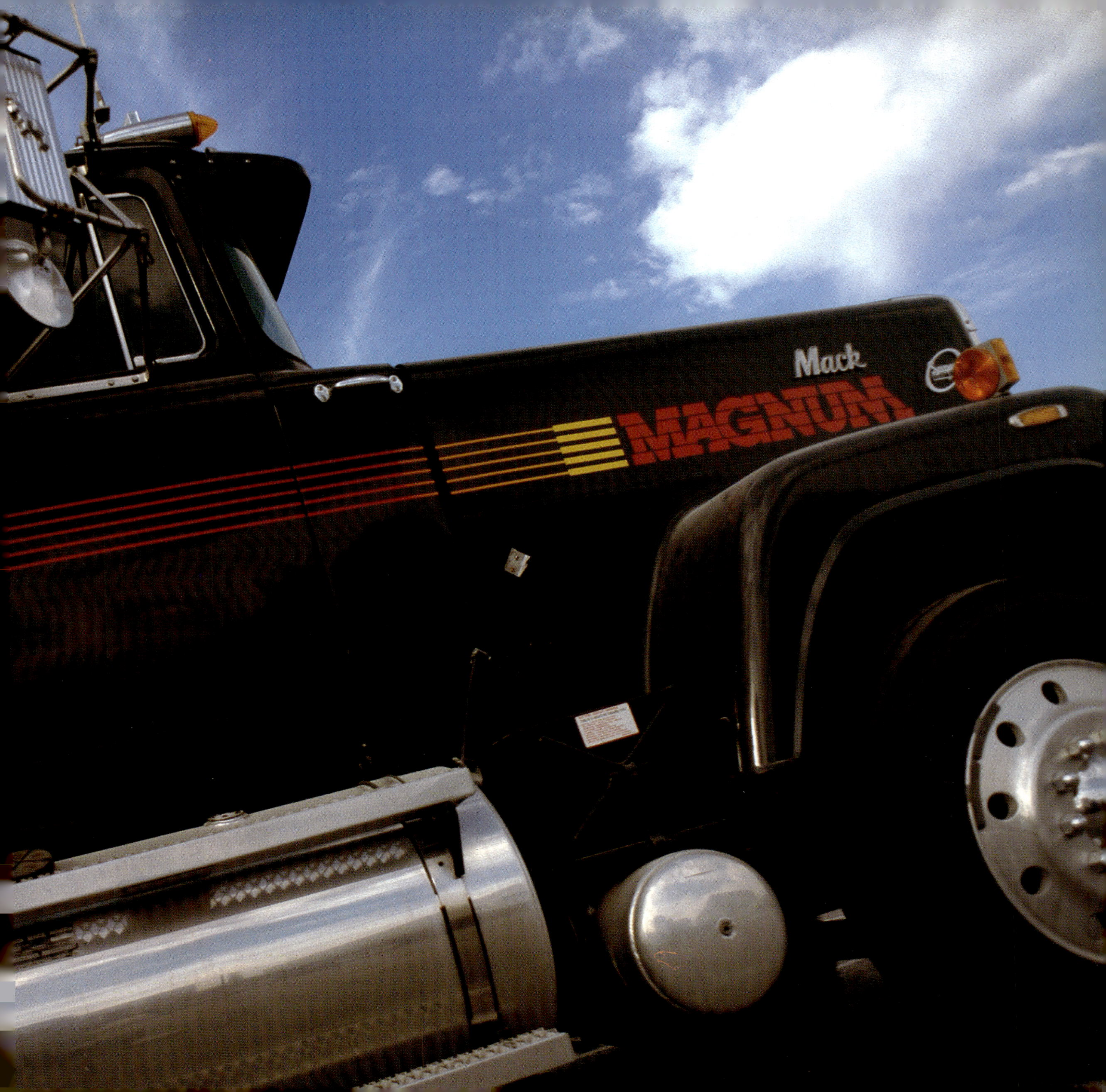
Mack
MAGNUM

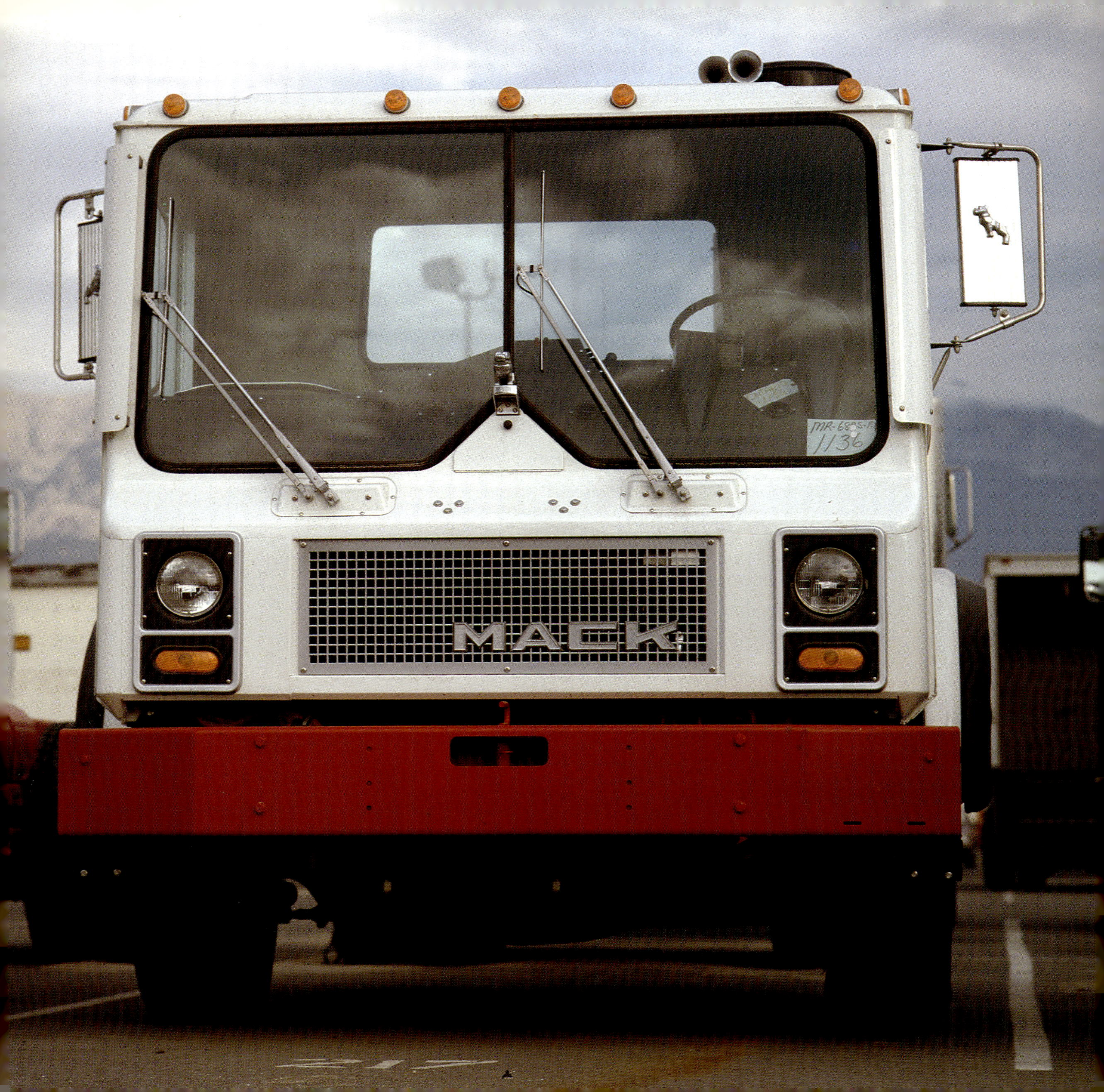
MACK

Featured on this spread are two faces of modern Macks. To the left, the MC, a heavy-duty urban delivery type truck, introduced in 1978. The stand-out feature of the styling of this modern Mack is the over-sized two-piece windshield, which tackles the problem of the increased need for visibility when maneuvering in tight spots and heavy traffic. Notice the enormous pantograph wipers. The cab tilts forward for easy engine access.

The photograph above shows the clean styling of Mack's new cabover, the Ultra-Liner, introduced in 1983. The sleek design was the result of extensive testing, both conducted in wind tunnels as well as arrived at with the aid of computers. In some markets, where the square lights do not conform to legal requirements, conventional, round headlights are fitted. This row of Ultra-Liners, its impressive geometry enhanced when multiplied, was captured in the storage yard behind Mack's truck center in Ontario, California, where they were awaiting final preparation and delivery.

Next page
The following spread features the newest face in the line-up of modern Macks. This handsomely styled conventional type truck, is also a Mid-Liner, the CS model, manufactured in France, by Renault. Like the cabover Mid-Liner, this conventional Mid-Liner is the result of a collaboration between engineers from both organizations, and they both carry a high degree of American-sourced parts. Power comes from a diesel, Renault-built, in-line six, the 335 ci version developing 175 hp, and the 538 ci version producing 205 hp.

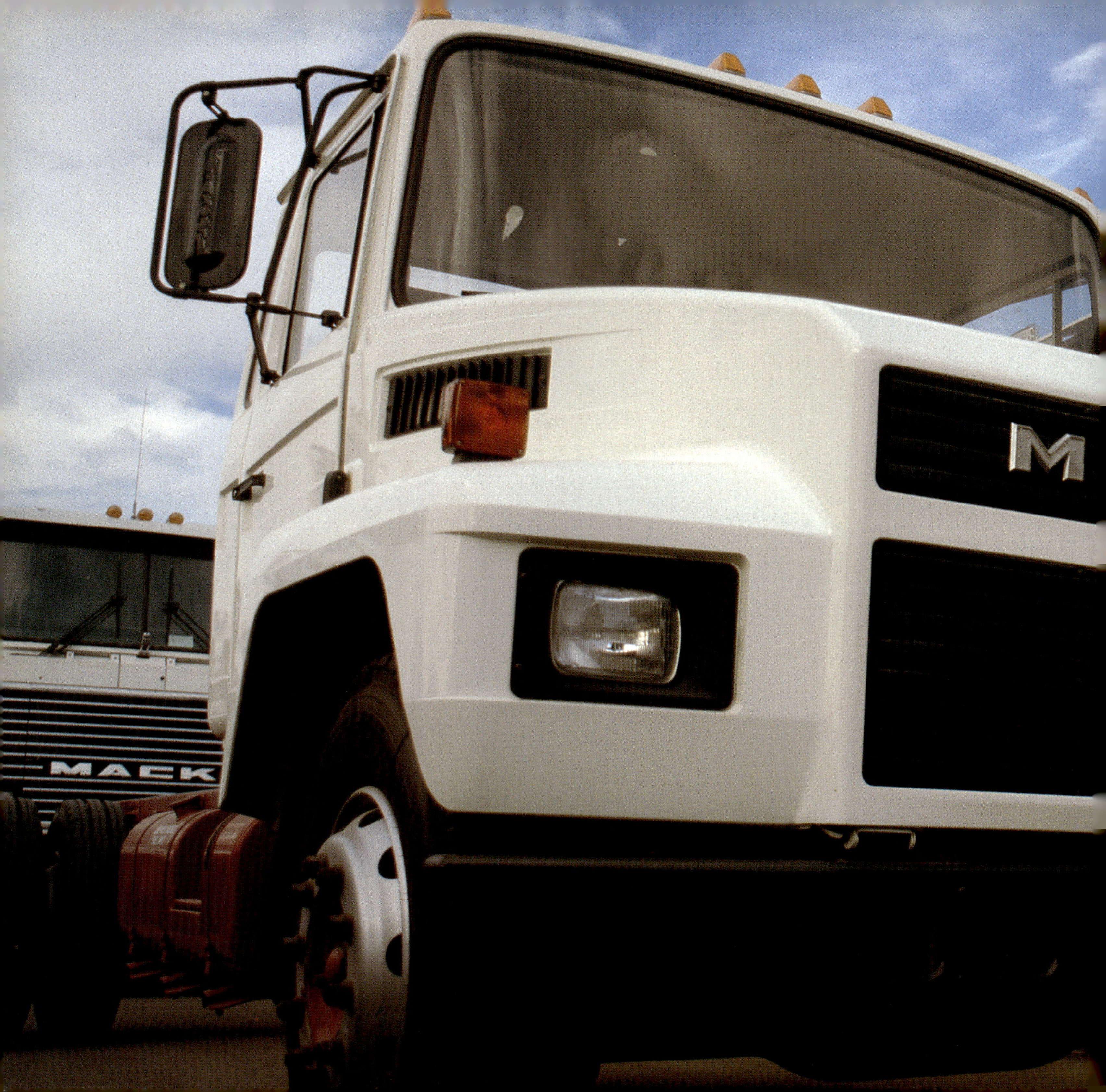
MACK
M

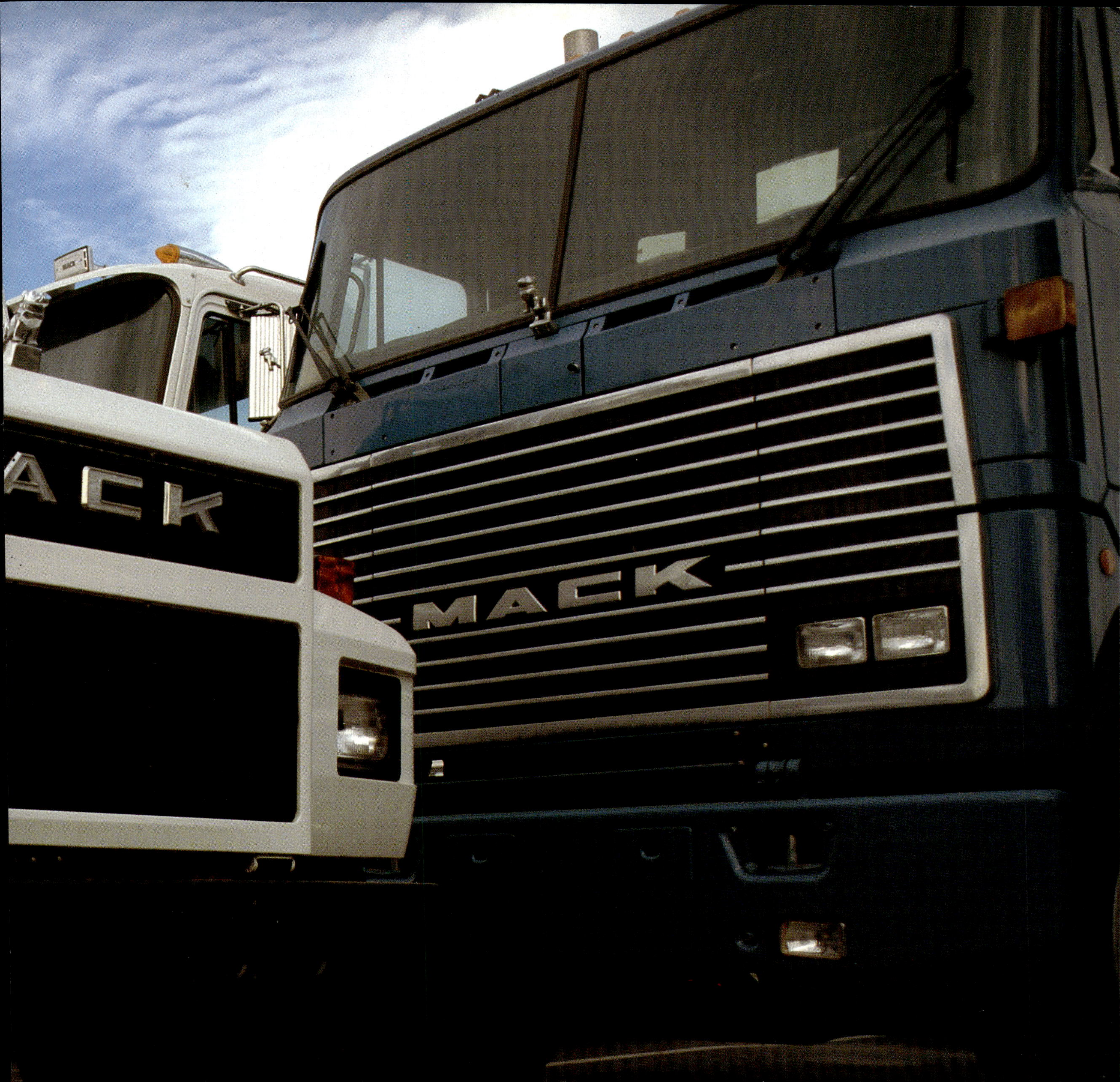
MACK
MACK

The photograph on this spread shows the rear aspect of an MR, the refuse—or garbage truck—version of the MC. The engine is Mack's six, which in the MR configuration can be had with power outputs of either 235 or 275 hp. Mack produces only diesel engines today, an in-line six or a V-8 configuration. All are turbocharged. The thick pipe protruding on the left side of the cab, is the air-cleaner unit. The turbo unit can be seen between the intake and exhaust pipes. The MR comes with Mack's own 6- or 7-speed transmissions.

Next page
The photograph on the following spread features a familiar old face. But here it has received a subtle new twist. For this is the Value-Liner, manufactured in the Macungie plant. The new machine looks very much like the old faithful R-600, which is still being manufactured in the Allentown plant. The Value-Liner is a lightweight version of this long-running veteran, and features Mack's 6-cylinder powerplant, with a 673 ci displacement and outputs varying between 250 and 350 hp. There is a choice of 7, 9 or 10-speed transmissions.

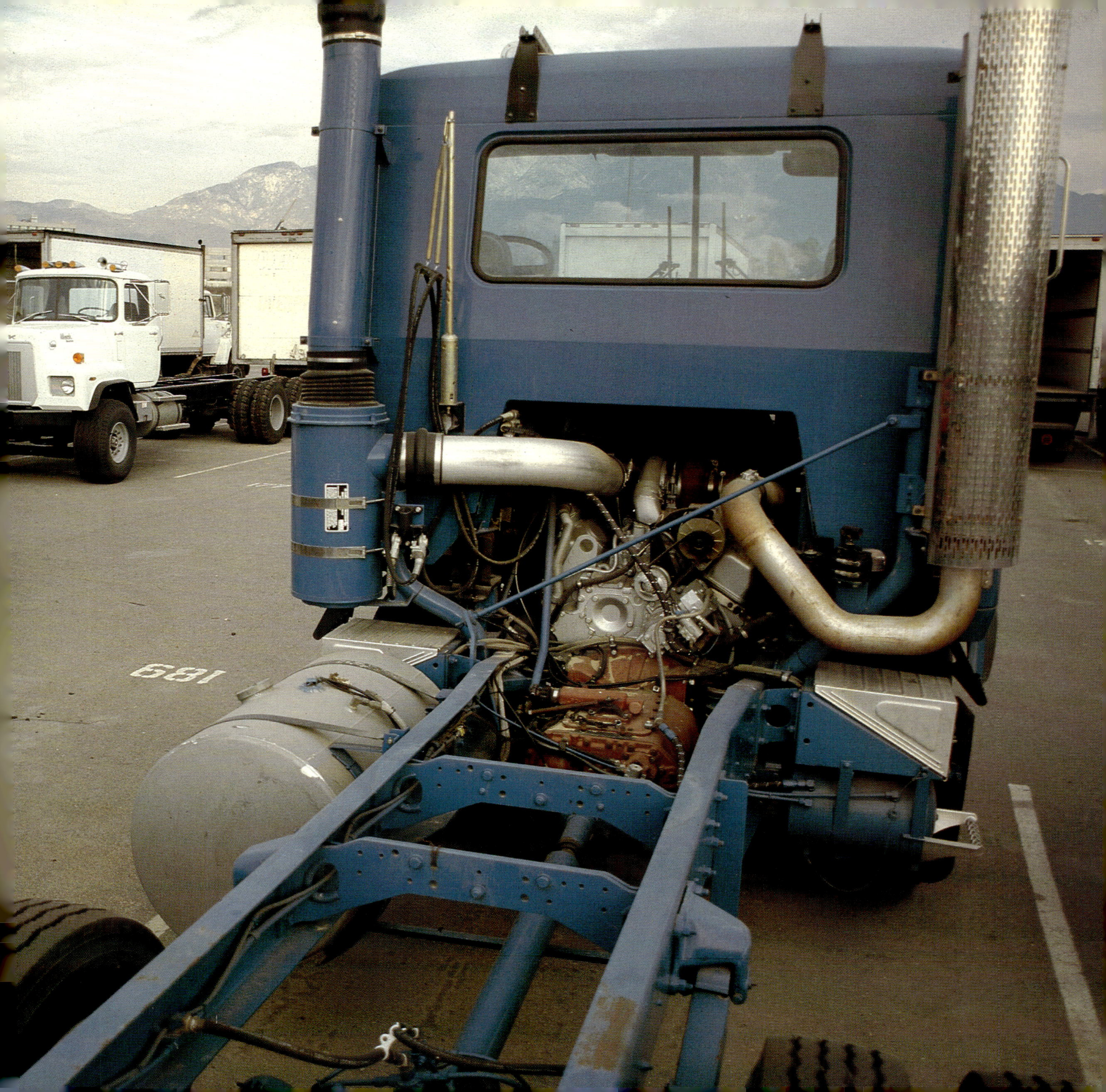

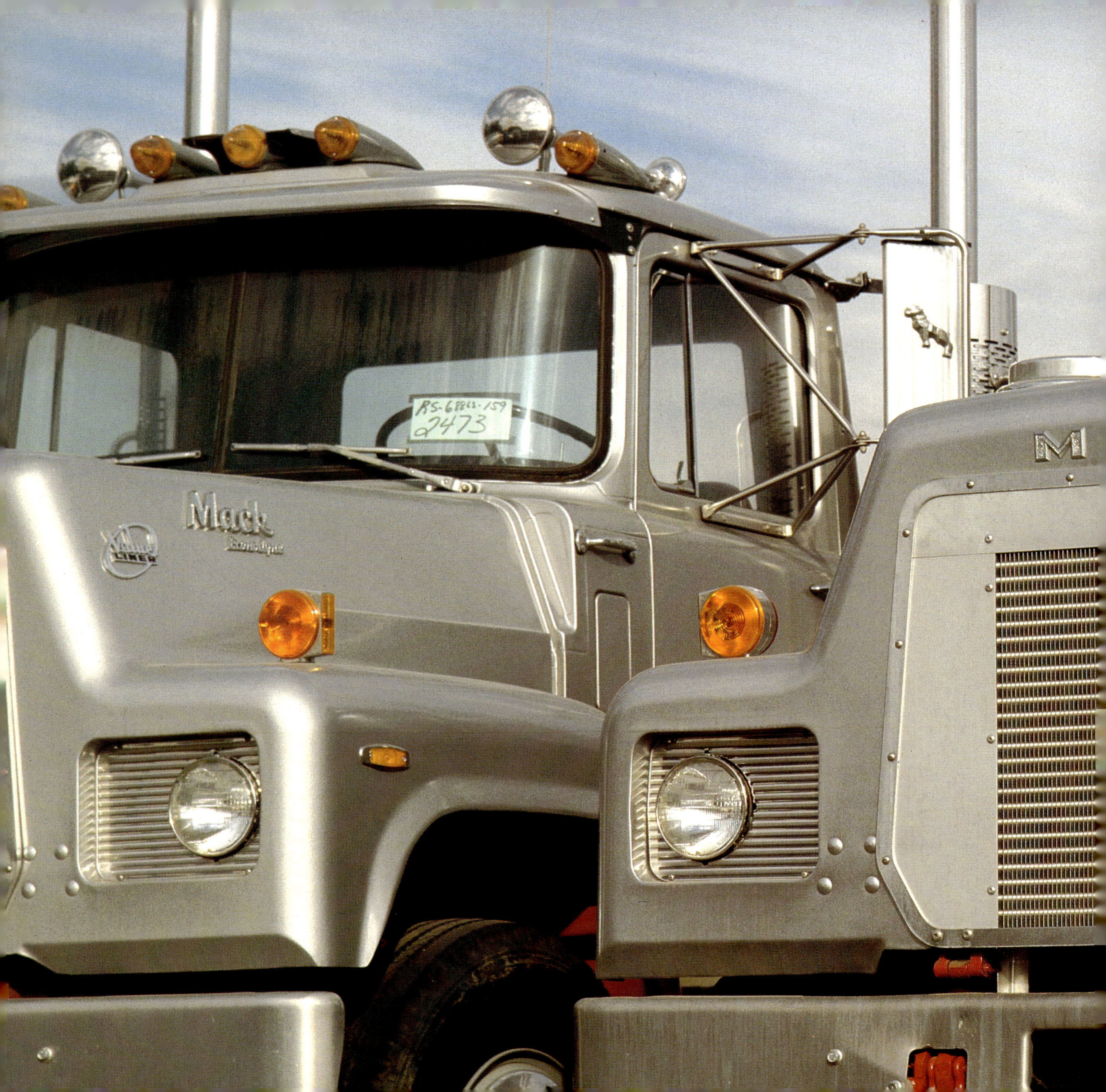
Mack
2473

RS 68844-H 40
2475

Mack Mania

Truck buffs show and tell

Trucks have been an important part of our every-day picture virtually since the turn of the century. The very few who can still remember that first decade, will have seen how the truck gradually took over from the horse, and those who remember the second decade, will have seen the truck grow to perform tasks the horse could never have accomplished. In the fifties, trucks and trucking made its final breakthrough, and most of us still remember those days with nostalgia, and some were even lucky enough to have played an active role in bringing it about. Today, old trucks are still around, gathering moss in foggy fields and hidden away in forgotten garages, but also displayed in museums, pampered in parades and admired in shows. The American Truck Historical Society has become one of the focusing points for all this love of trucks, with members spread all across the nation and chapters in major cities, as well as an annual show of national scope. Mack, with its pioneering role, naturally takes a front seat in any such event. The following pages show glimpses from an ATHS gathering in Seattle, Washington.

A perennial favorite among Mack enthusiasts is the legendary LTL, a 1953 example of which is seen in the photograph on this spread. The Mack classic, with its long hood and wide grille, combines Mack power and prestige in a most appealing way. Here, its massive chromed bumper mirrors a line of fellow show companions, of which the two red ones are Macks.

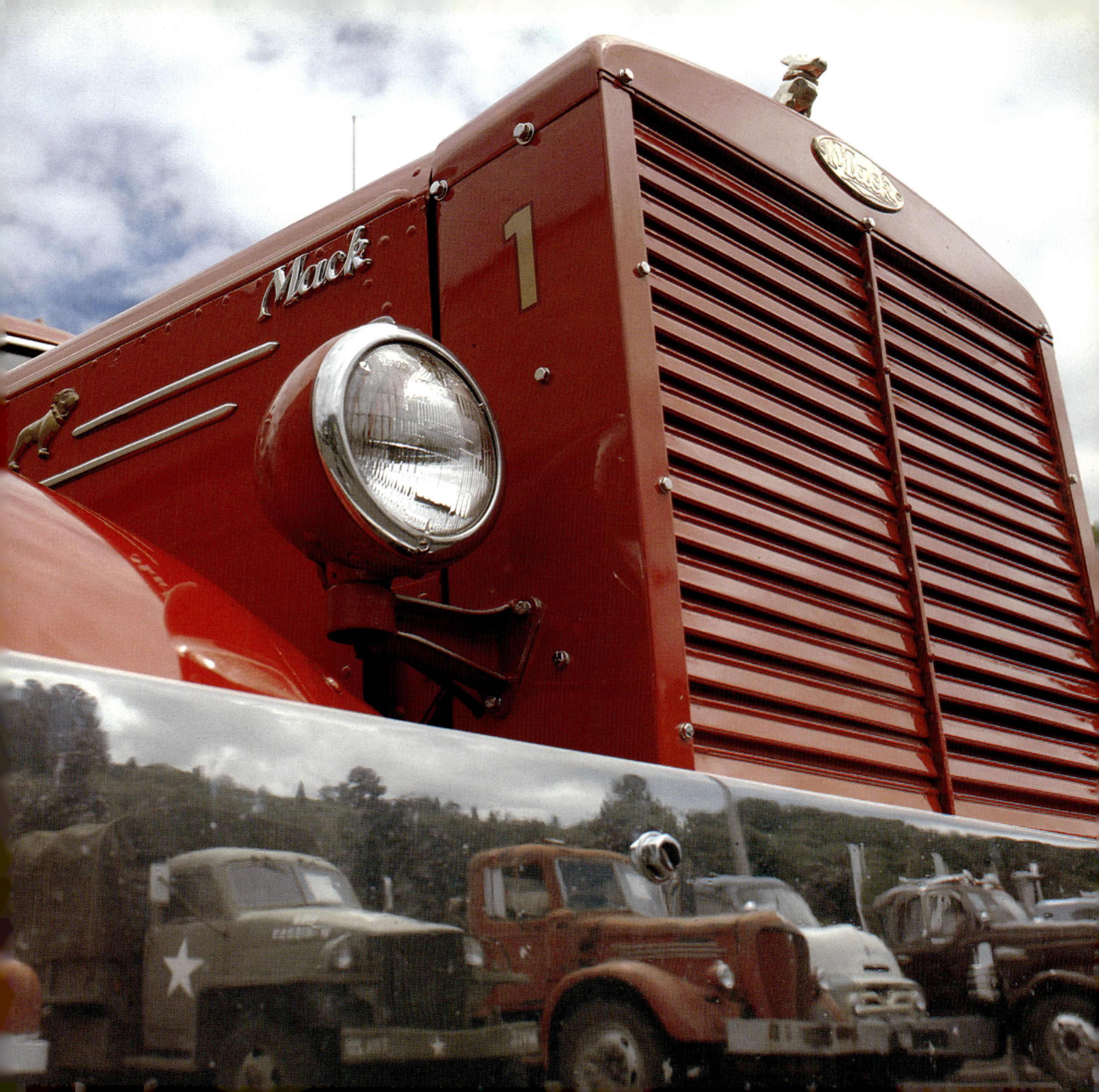
Mack
1
Mack

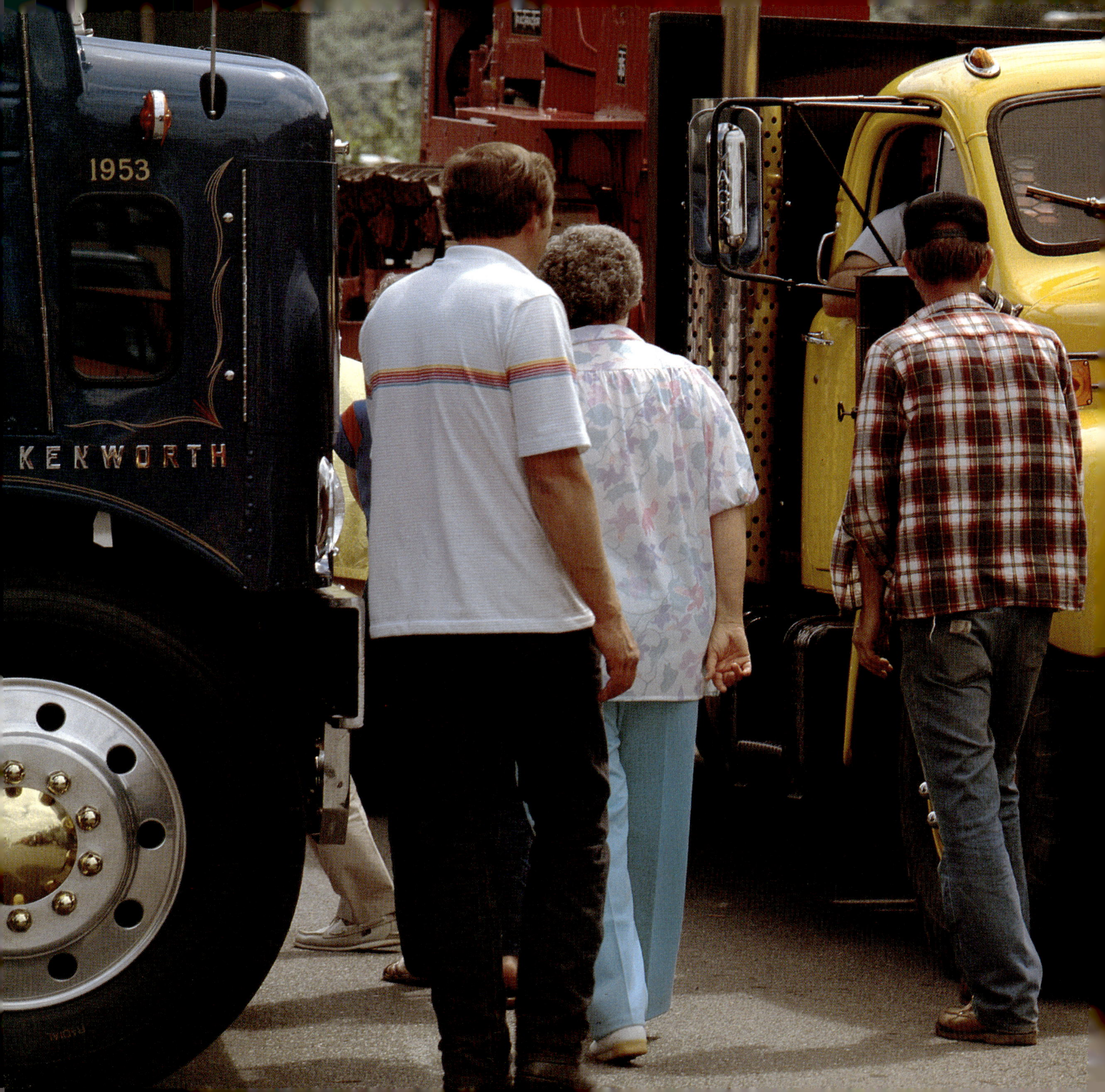
1953
KENWORTH

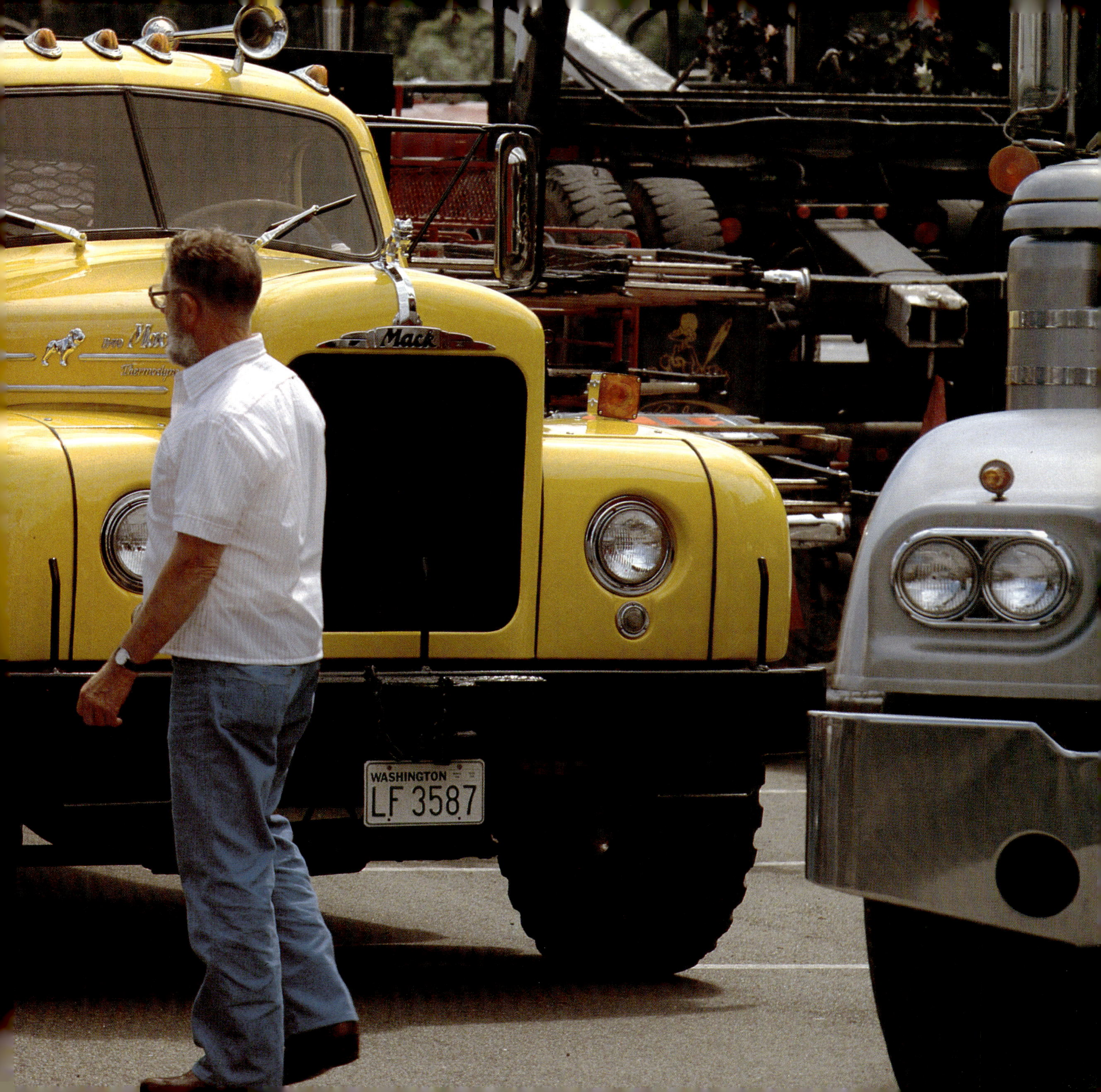
Mack
WASHINGTON
LF 3587

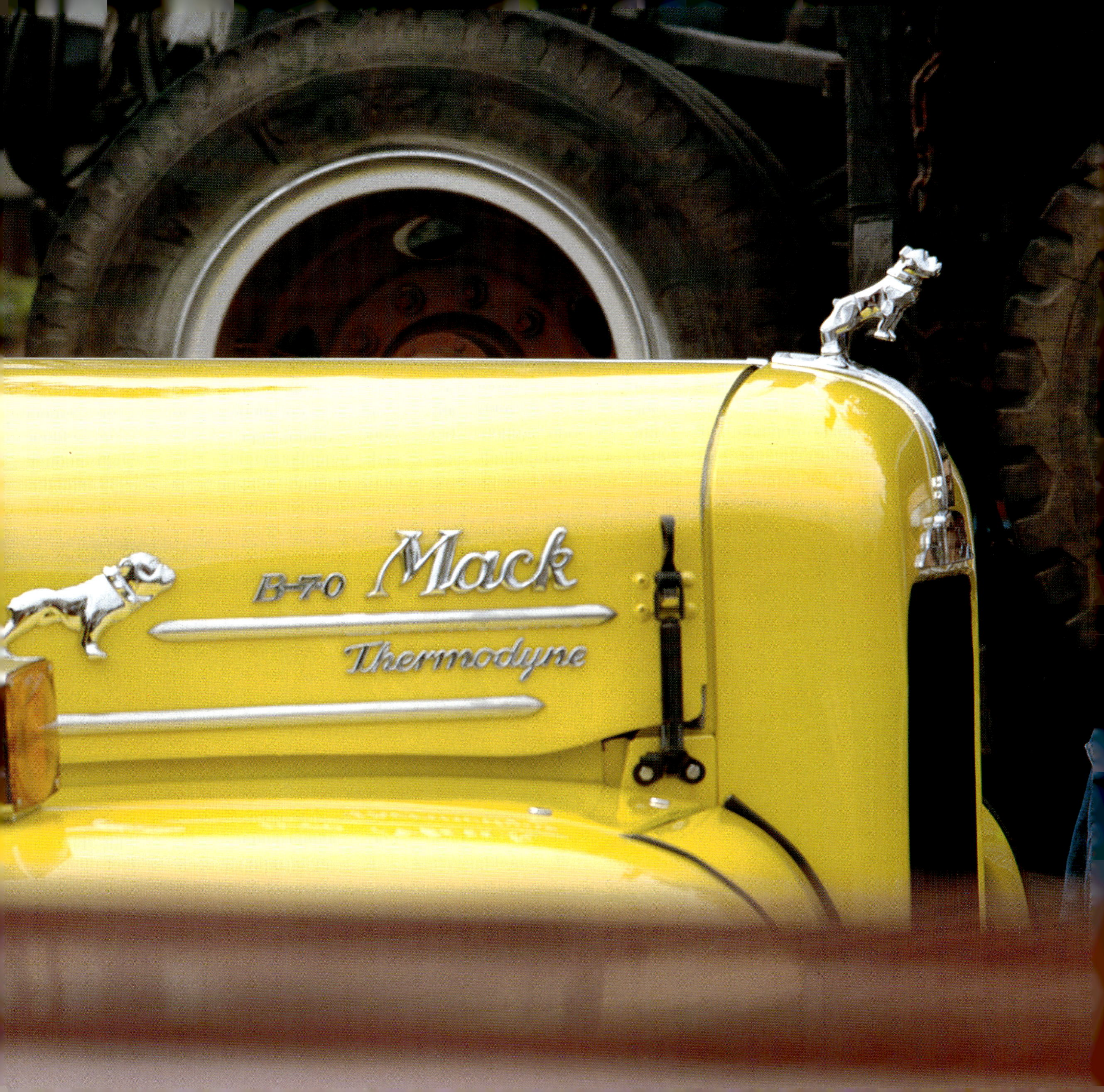
B-70
Mack
Thermodyne

Previous page

The Mack featured in the photograph on the previous spread is seen on this spread as well, proudly displaying its flawlessly chromed bulldog mascot—a symbol recognized and admired by old and young alike. Truck collecting is still something of a virgin territory, having not yet reached the sophistication of car collecting, where certain aspects of that hobby sometimes seem to have passed into the realm of the business world.

The ATHS annual events are usually held at a convention-type hotel, the size of its parking lot being one of the most important aspects of the selection criteria, since the members' trucks—hundreds of them—stay there for the three-day duration of the gathering and doubles as a show arena. In the photograph on the previous spread, a B-70 Mack arrives late, and already before having managed to locate its spot, the beauty, with its eye-catching shape and color, begins to attract a crowd.

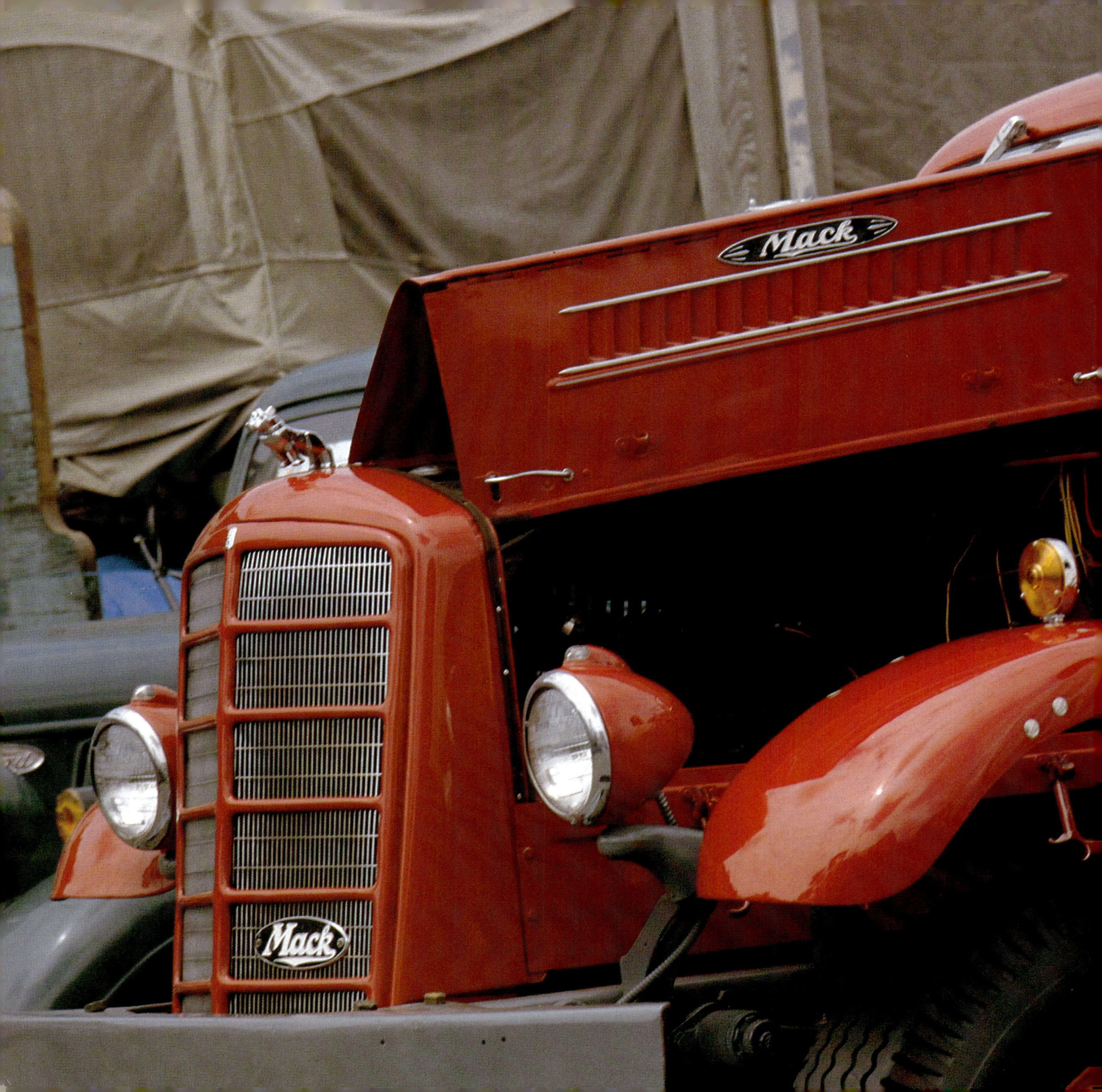
Mack
Mack

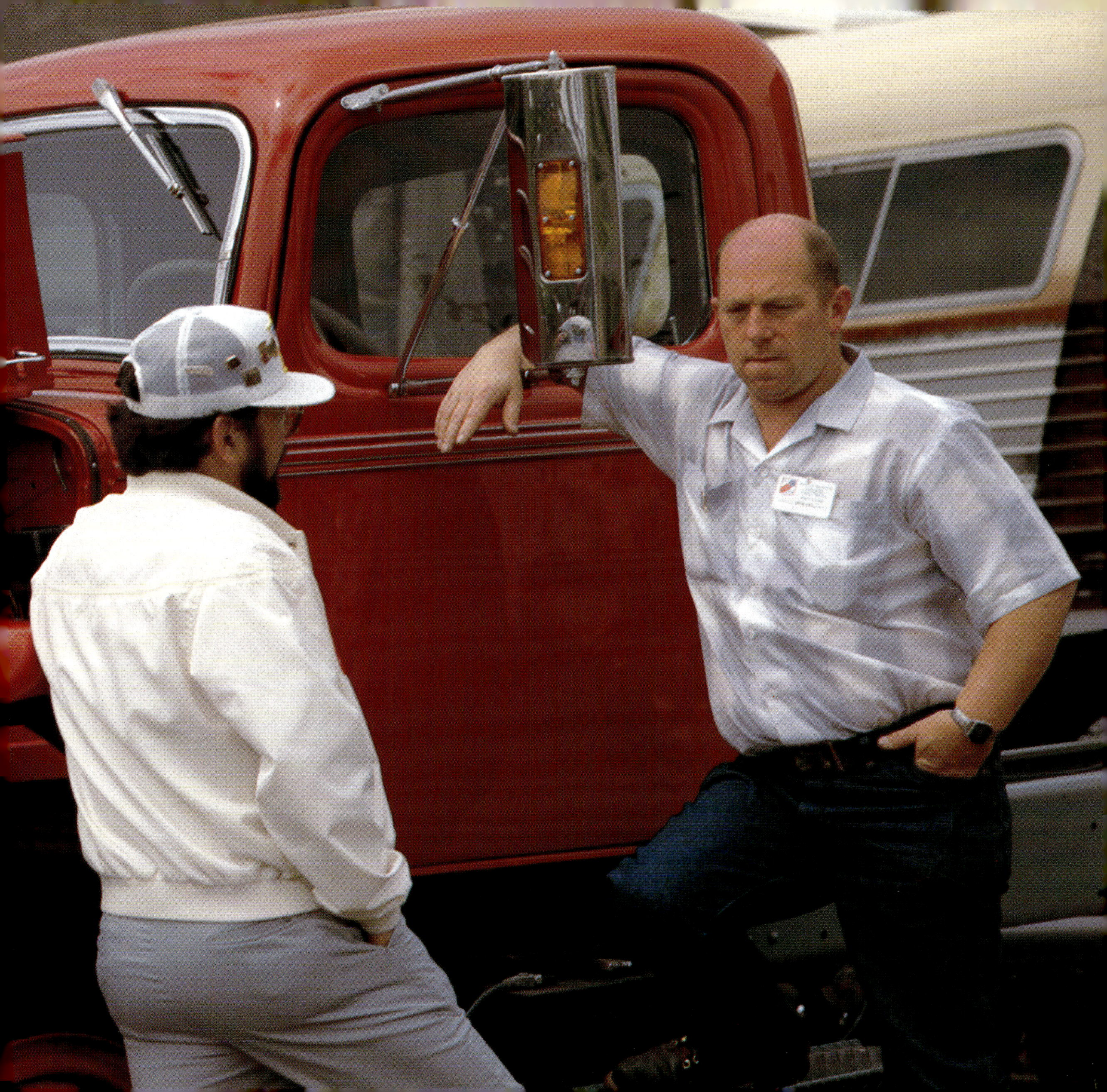

Previous page
The annual ATHS conventions give members a chance to show and tell. In the photograph on the previous spread, Mack enthusiast Bruce Berg from Bellingham, Washington, explains the delicacies of his unique 1949 EQX, an off-road vehicle belonging to the E-series, which was introduced in 1936. The exquisitely restored Mack, with its unusual cycle-type fenders, was originally fitted with a construction crane.

Members of the ATHS gather from all corners of the nation, sometimes traveling behind the wheel of their own priceless collector pieces. One enthusiast drove his 1921 Mack AC from Bakersfield, California. Another faithful, Joe Becker, pictured in the photograph above, drove his 1951 LF, a sub-model of the L-series introduced in 1940, all the way from Albert Lea, Minnesota.

Above, Becker shows the engine compartment of his LF, occupied by Mack's first diesel, the Mack-Lanova. This now hard-to-find engine had a 519 ci displacement, and produced 131 hp. The photograph to the right shows a full-face view of Becker's pride and joy. Coupled to the tractor is the Winnebago that serves as Becker's home during the trip. The equipage is quite a sight at speed, with the bulldog showing the way and the Stars and Stripes flying in its wake.

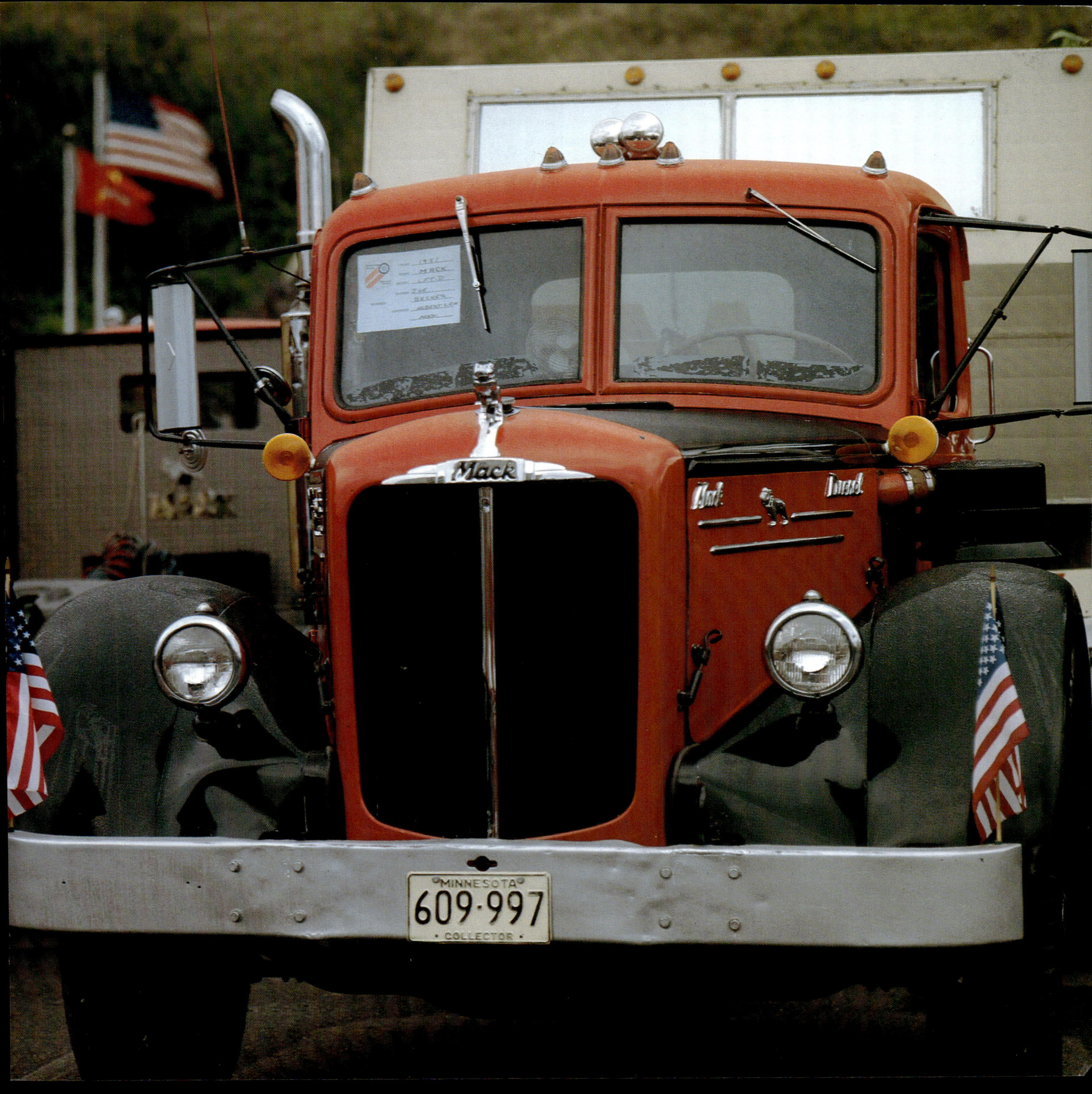
Mack
MINNESOTA
609·997
COLLECTOR

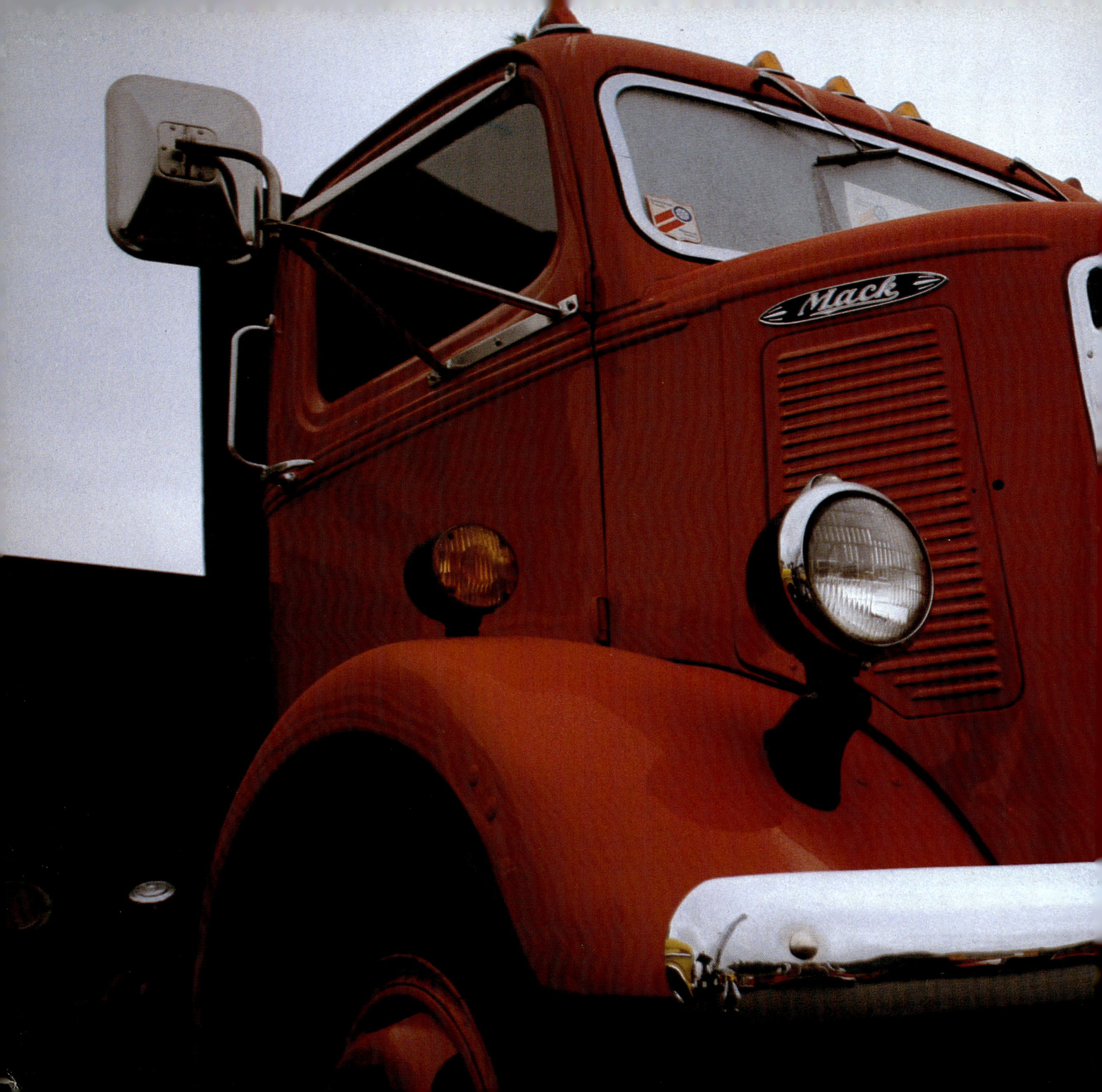
Mack

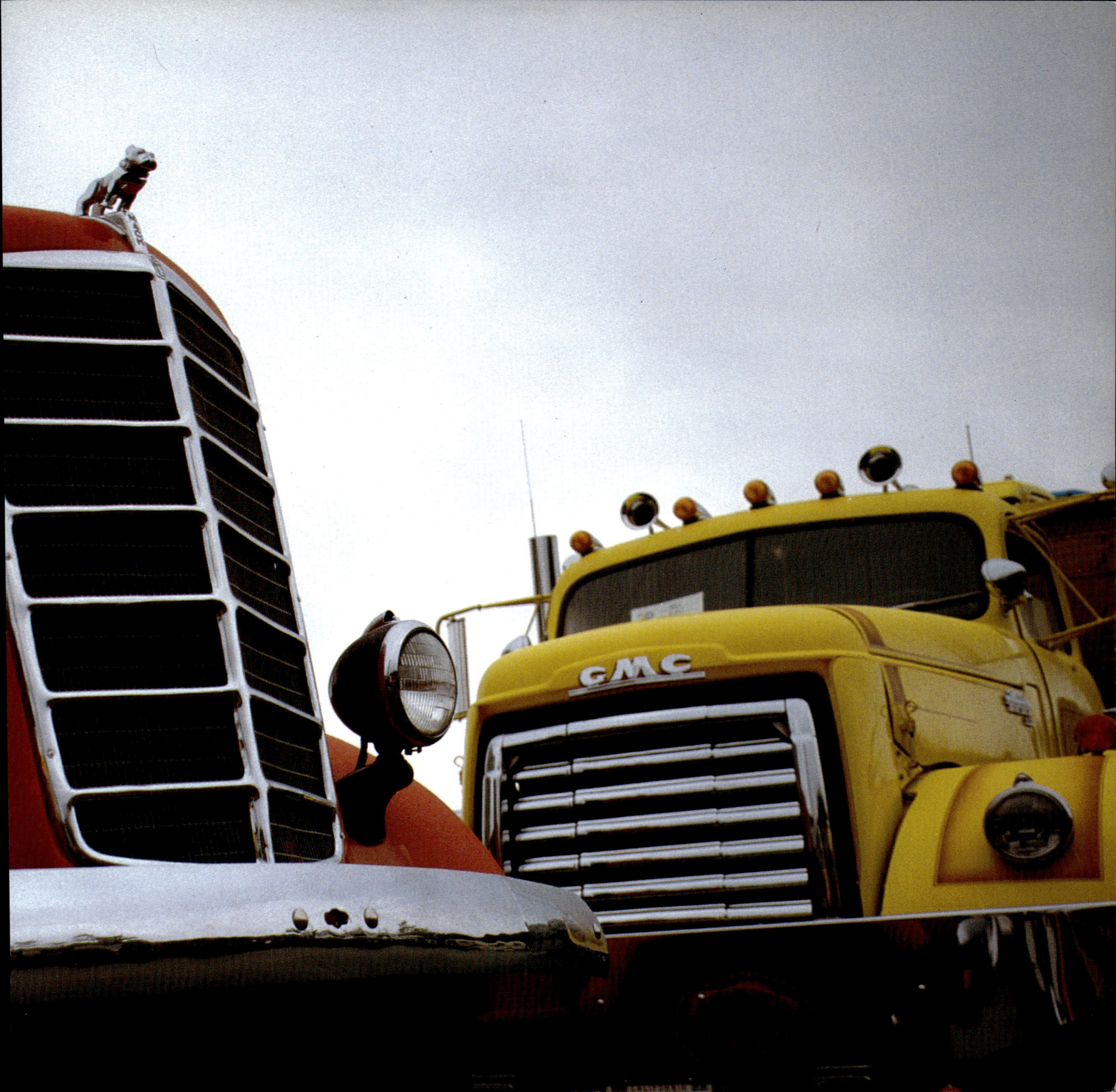
GMC

Previous page
Early after-rain mornings often offer moody photo opportunities, as in the picture on the previous spread, featuring a 1940 EH Mack, owned by Gary Klokstad, of Lynwood, Washington. The E-series, which was introduced in 1936, was the first to feature modern, streamlined styling. The EH was powered by Mack's 79 hp BG-type engine. A 5-speed transmission with overdrive was available as an option. A total of 31,539 units had been manufactured by the end of production in 1950.

Pictured on this spread is a unique Mack, a 1953 W-71, owned by William Comcowich, of Aspen, Colorado. Power for the cabover giant comes from a 200 hp Cummins diesel, and the overall height of the cab is an impressive 9 feet. The W-71 was introduced as a companion to the conventional LTL, which both were—with their lightweight construction—especially designed for West Coast operation. Just 215 units were built between 1953 and 1958.

Even though the hey-days of this fifties' Mack lay beyond his sphere of memory, the young man in the picture to the left is still fascinated. The historic significance of the beautifully restored survivor is of little consequence to him. It is the bold surfaces and the shiny chrome and—most of all—the sheer size, that fills him with awe. Who would not be impressed by a vehicle if its wheels alone reached almost above one's head?

The inclusion of the photograph above should by no means be misconstrued as a chauvinistic comment, alluding to the possibility that the members of one particular sex may show less interest in things that have to do with trucks and trucking, but it is simply an attempt to state the fact that, no matter how nice and clean and shiny and big everybody else's wheels are, the comfortable security of one's own wheels are always to be preferred.

Trucks and the life of the trucker has always been thought of in romantic terms. To most truckers, however, once the honeymoon is over, it is just a job like so many other jobs, maybe a little tougher than most, and lonelier. But once in a while, perhaps when seeing a sight like the one above—early morning, empty roads, plenty of time to think— the old romance is back.